AF885573

Naturally Postnatural

Catalyst Book Series

01 Plastic Blue Marble—Catalyst: Amanda Boetzkes

02 Phono-Fictions and Other Felt Thoughts—Catalyst: Eldritch Priest

03 Naturally Postnatural—Catalyst: Jennifer Willet

Series editors

Amanda Boetzkes
David Cecchetto
Ted Hiebert

Naturally Postnatural

Catalyst: Jennifer Willet

Edited by Ted Hiebert

Contributors

Warren Cariou
Louise Chance-Baxter& and IAIN BAXTER&
George Gessert
George Gessert & Beth Franks
Christian Kuras
Marta de Menezes
Natasha Myers
Kira O'Reilly
Melentie Pandilovski
Paul Vanouse
Amanda White & Alana Bartol
Robert Zwijnenberg

Noxious Sector Press
Victoria • Seattle

© 2017 Noxious Sector Press

All rights reserved. No part of this book may be used or reproduced without written permission except in the case of brief quotations embedded in critical articles or reviews. All essays and artworks are copyright by the authors.

Editor: Ted Hiebert
Contributors: Warren Cariou, Louise Chance-Baxter& and IAIN BAXTER&, George Gessert, George Gessert & Beth Franks, Christian Kuras, Marta de Menezes, Natasha Myers, Kira O'Reilly, Melentie Pandilovski, Paul Vanouse, Amanda White & Alana Bartol, Robert Zwijnenberg.

Cover image: Jennifer Willet & Kira O'Reilly, *Be-Wildering*, 2017. Color photograph, photo credit: Bas de Brouwer.

Noxious Sector Press
noxious@noxioussector.net
http://www.noxioussector.net

Library and Archives Canada Cataloguing in Publication

Naturally postnatural : catalyst : Jennifer Willet / edited by Ted Hiebert ; contributors, Warren Cariou, Louise Chance-Baxter& and IAIN BAXTER&, George Gessert, George Gessert & Beth Franks, Christian Kuras, Marta de Menezes, Natasha Myers, Kira O'Reilly, Melentie Pandilovski, Paul Vanouse, Amanda White & Alana Bartol, Robert Zwijnenberg.

(Catalyst book series ; 3)
ISBN 978-0-9936058-3-3 (softcover)

1. Willet, Jennifer, 1975- --Criticism and interpretation. 2. Art and science. 3. Essays. I. Hiebert, Ted, 1973-, editor II. Title: Catalyst : Jennifer Willet. III. Title: Jennifer Willet. IV. Series: Catalyst book series ; 3

N72.S3N38 2017 701'.05 C2017-905776-6

Catalyst

The twenty-first century is a time of prodigious creative and intellectual experimentation, with many thinkers, artists, and makers engaging in a range of practices that are foundationally speculative yet nevertheless transformative. The Catalyst book series aims to represent this space of possibility by coupling theorists and artists in ways that galvanize logics, spaces, politics, and practices that are not yet mapped … and perhaps never can be.

Catalysis instigates processual differentiations over a space of exchange; it is eventful, unpredictable, and generative. To chart a catalyst is to bring attention to the critical and creative processes that reveal hidden perspectives upon the event of their becoming. Thus, contributors to the Catalyst books think *alongside* the catalyst, edging and forging implications, connections, atmospheres and weirdnesses. The essays do not review or critique the catalyst's work but rather sound points of contact in pursuit of resonances, enacting gestures of performative solidarity through intellectual and creative engagement.

Catalyst books build speculative communities, inviting a wide range of perspectives into conversations about shared artistic, political, and intellectual values while privileging the unique, distinct and personal insights that characterize any single voice of engagement. Each volume in the series provides an in-depth look at an active thinker or artist—seeking after the full relevance of their work. The series focuses in particular on voices that have not already been widely featured but who have unique and relevant perspectives to share on questions of art, theory and culture.

Contents

Catalyst: Jennifer Willet

Jennifer Willet is an artist, researcher, performer, and curator in the international art/science community. She is an innovator in the field of bioart, merging artistic and biotechnological research towards research/creation that uses living media (cells, plants, microbes) in the production of artworks.

Willet began her career as a founding member of the artist collective BIOTEKNICA (in collaboration with Jason Knight). BIOTEKNICA was a fictitious bioengineering corporation that investigated the ethics, aesthetics, and technological potential of bioArt practices. Since 2008, Willet has collaborated with UK/Finland based artist Kira O'Reilly conducting unruly photoshoots in laboratories all over the world. This work has been presented widely, most recently as part of the "Trust Me I'm an Artist" public ethics review series co-curated by Anna Dumitriu and Lucas Evers at the WAAG Society in Amsterdam.

Willet also has a thriving individual art practice that explores notions of performing bioethics, laboratory ecologies, and interspecies interrelations in biotechnological fields. New work includes a series of portable lab equipment/furniture sculptures, and a bicycle propelled street organ that grows and displays living spirulina algae. Audiences are invited to enjoy live organ music while engaging with a live algae colony. The series also includes a functional biological safety cabinet that is designed to look like a miniature theatre set, and a functional centrifuge upholstered into a vibrating antique settee.

In September 2009, Willet launched the first bioart laboratory in Canada: INCUBATOR Hybrid Laboratory at the Intersection of Art, Science and Ecology. In 2011, INCUBATOR Lab and The Banff Centre co–produced a two-week artist residency called BioARTCAMP in which Willet hosted 20 artists, scientists, and students camping and building a portable bioart laboratory in Banff National Park. The results were presented in a traveling exhibition called NATURAL SCIENCE, a collection of hundreds of items, life forms, images, and stories resulting from the BioARTCAMP project.

In 2018, INCUBATOR Lab will launch a new biosafety level 2 bioart lab facility. This state-of-the-art hybrid laboratory/multimedia theatre, will host international bioart events where live and online audiences will be able to experience hands-on biotechnological protocols embedded into multimedia performance events.

Willet grew up in Calgary, Canada. She holds a BFA from the University of Calgary (1997), an MFA from the University of Guelph (1999), and a PhD from the Interdisciplinary Humanities Program at Concordia University (2009). She is an Associate Professor in the School of Creative Arts at the University of Windsor. In 2017 Willet was awarded membership in the College of New Scholars, Artists, and Scientists of the Royal Society of Canada.

Jennifer Willet lives and works in Windsor, Ontario with her partner Grant Yocom and their two children. http://www.jenniferwillet.com/

Introduction

Ted Hiebert

The idea of nature in the 21st century is marked not only by the concession that humans have had an impact on a global environmental scale, but also by the fact that a humanly-impacted environment is now the default—one might say natural—state of planetary affairs. The twin markers of this discourse are the geological assertion that we have entered the age of the Anthropocene—"Earth in the age of the human," as Julie Hannon puts it—and the growing critique of anthropocentrism as more than a casual hubris of historical thought.[1] To state the obvious would be to say that anthropocentric thinking is responsible for the Anthropocene, proving once and for all that ideologies have environmental impact, even on a geological and planetary scale.[2] The critique of anthropocentrism goes further than this, of course, seeking out non-human (or non-human-centric) ways of thinking and engaging with the world in hopes of finding reparative alternatives to the manifest ideologies of the past. And it comes at a good time too, since everywhere technological advances seem to be intensifying the materialization of thought, information, and the imaginary in new and sometimes surprising ways. We live in an era where cloning is old news, self-driving cars and drones are becoming commonplace, and artificial intelligence has been granted citizenship.[3] It often gets called posthuman, but if it is posthuman it is also decidedly postnatural, since thought beyond these categories will inevitably have to understand itself as already entangled with its capacity to adapt to life without the (historically) usual boundaries. Nowhere is this more pronounced than in biotechnological contexts where the boundaries between the human, the engineered, and the natural have long since been reimagined as variations in a common language of genetic code, transferable beyond species boundaries. Humans can now receive animal transplant organs, organisms of all kinds are regularly grown in laboratories around the world, and just this year human brain cells were inserted into the neural cortex of a rat.[4] This is thought and technology applied well beyond the constraints of what used to be natural. This is the naturally postnatural.

This could all be put a bit differently.

Joseph Beuys once famously declared that "everyone is an artist" and that together we are building the artwork that will be our collective future. According to Beuys, teachers are artists (creating ideological patterns in the minds of their students), parents are artists (literally creating new forms of life), and politicians are artists (working to shape—even sculpt—the social and economic worlds in which we live). And factory workers are artists too, as are gardeners, nuclear physicists, and hooligans. What all modes of life have in common—for Beuys—is that the actions taken by individuals, whatever the professional or personal particulars, can be seen as a form of "social sculpture," the creation of a public stage on which life as we know and encounter it is to be lived.[5] The simple act of living impacts the stage upon which life is more broadly lived.

It would not be an exaggeration to say that the Anthropocene is the unexpected consequence of lived human engagement, and as such one might argue—following Beuys—that it is an artwork of geological proportions.[6] Nor are we easily separable from the work itself. If the Anthropocene is an artwork, it is one that we are all living within, a "theater of cruelty" in which there is really no difference between artists, actors and audience, where in fact artists don't always even know themselves as such and actors forget that they are reading scripted lines. It is a catalytic state, though it is never quite clear where the catalyst ends and the system upon which it acts begins. To make this claim is to insist, simply, that living is biocatalytic: a transformation specifically linked to living systems. Biocatalysis is an acknowledgment that catalytic change has real effects, material effects—reminding us that there is no metaphor involved. We are as much the medium of change as we are its subjects.

Catalyst Jennifer Willet

Catalyzing ideas of this sort requires a commitment to both analytic method and creative intervention, an artistic spirit with an ecological conscience. And there is no better figure for such an exploration than Jennifer Willet, whose practice as an artist, an educator, and a social provocateur can be so catalytic because she herself is constantly engaging new forms of artistic interaction. Sometimes she creates, sometimes she simulates, sometimes she organizes, curates, or performs—always with a concerted insistence on the ultimate social value of discourse. For Willet, discourse is not rhetorical however: it is biological—and the stage upon which her work is situated is more concerned with dialogue and community than with argument. In this, Willet's work is decidedly postnatural, mobilizing creative forms for their dialogic potential, and in so doing demonstrating how the (artistic) naturalization of postnatural forms can serve as catalysts for social, creative, and philosophical discourse.

Willet identifies first and foremost as a bioartist: an artist working with biological materials and processes as her

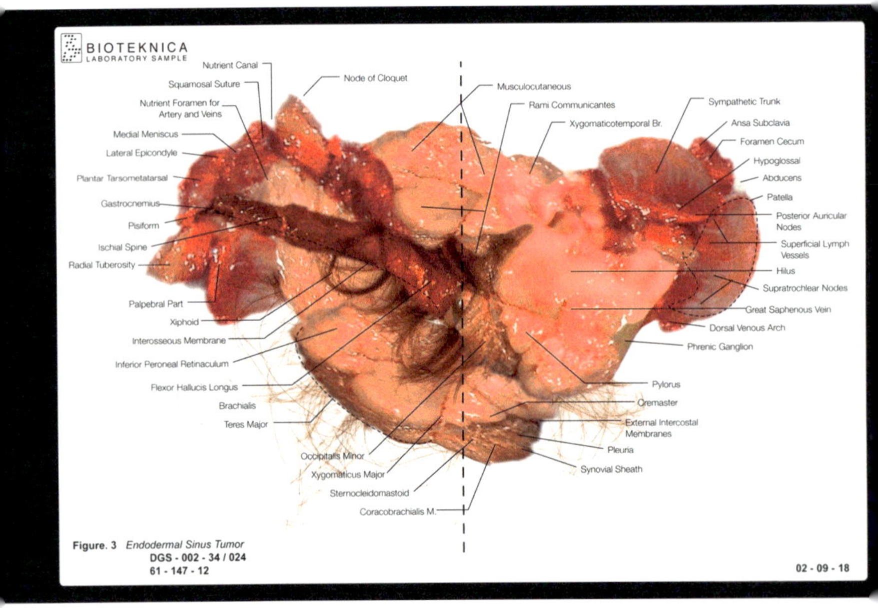

Jason Knight & Jennifer Willet, *BIOTEKNICA: Teratoma Anatomy*, 2004. Digitally generated duratrans.

medium. It is important to note that, as a medium, bioartistic practice is already premised on a postnatural ideology, one that accepts and embraces the possibility of manipulating the natural to artistic ends. BIOTEKNICA (2001-07) is an early instance of this method, a project in which Willet and her collaborator marketed digitally-simulated teratomas, creating a fictitious corporation around the idea of genetic marketing and the economic potentials of new forms of life. When working at the limits of natural possibility, the reality of a situation can often be less important than its plausibility—and the revelation of BIOTEKNICA is not the subterfuge, but the proposition for cosmetic biodesign: terratomas as family pets, design features of a new form of urban living.[7]

Importantly, it is a proposition to which Willet holds herself accountable as well and is particularly visible in performative works such as her collaborations with UK performance artist Kira O'Reilly. Making work premised on the interrogation of laboratory spaces as sterile or neutral environments, Willet and O'Reilly insert their own bodies—often nude—into these sites as a gesture of creativity, fertility, contamination, and play. These actions become even more performative in newer works in which stoic poses, costumes, and dramatic settings form the artistic locus of engagement. Elaborately attired, elegantly posed, impeccably positioned within the frames of European architectural structures, this is a theater of wearable science. And on this stage, Willet's own body forms the structural architecture for laboratory growths, each contained in their own tiny biodome sewn to the fabric of her costume.

One thing about Willet's work that is important to note is that it is also often reversible—the costume she

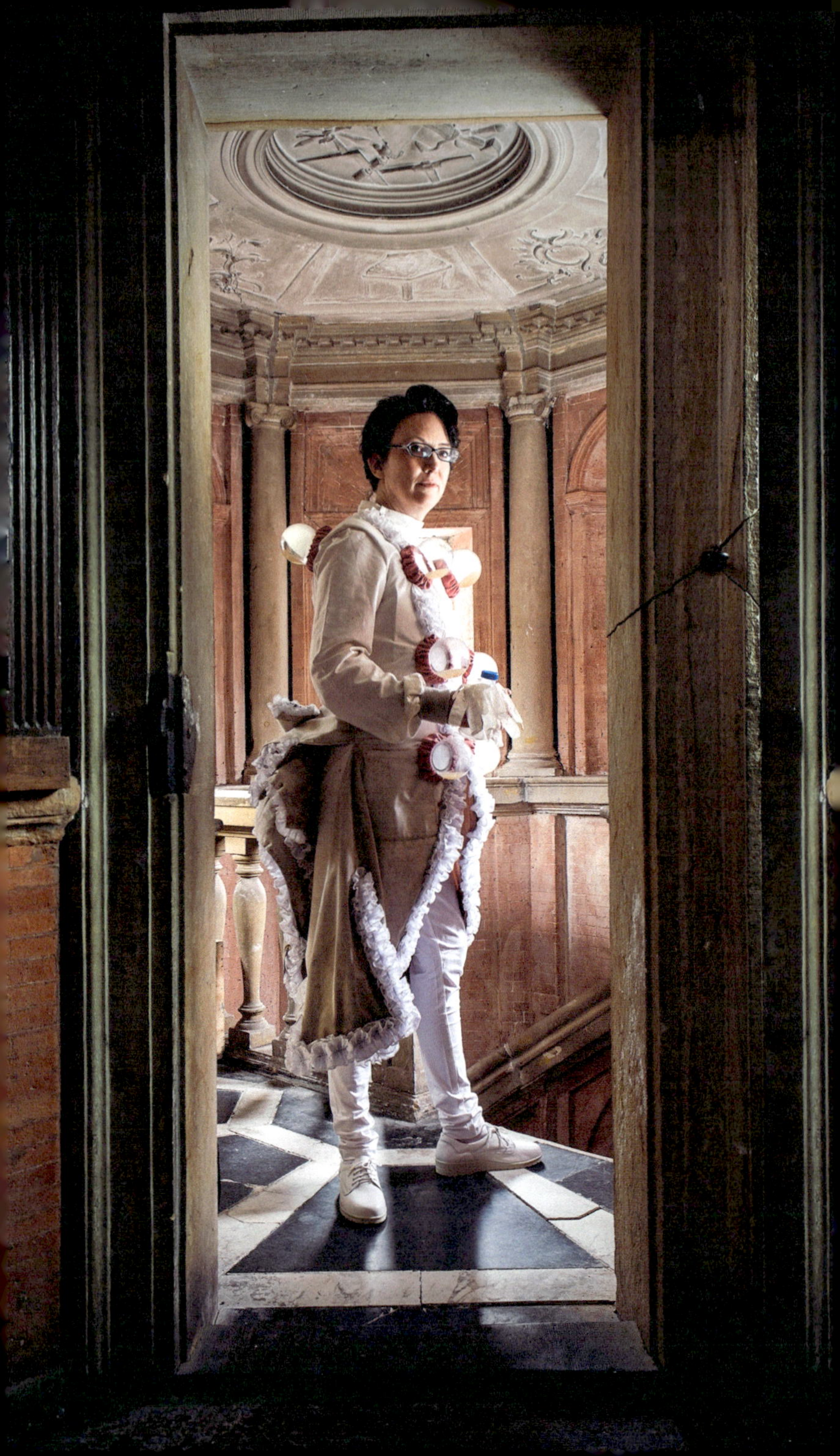

wears is a metaphor for our own fashioned engagements, and her simulations of biodesign are interrogations of ideologies with which we too comply. Her works are inside-out, which—for Willet—means that personal interaction is often a metaphor for social engagement. Like Beuys, Willet's work becomes its own proposition for a form of social sculpture—sometimes metaphoric, but just as often literal. Her *BioARTCAMP* project (2011), for instance, dedicates itself to the creation of a platform for others, an artistic retreat in the heart of the Canadian Rockies where artists and scientists spent two weeks living and working together, creatively and collaboratively exploring the interface of art and life.[8] In Willet's words: "*BioARTCAMP* is designed to emphasize ecological metaphors for describing biotechnology in public discourse and to complicate the 'Great Divide' between lab and field based research methodologies in the hard sciences."[9] Using a deeply natural setting as a stage, this project is nothing if not collaboratively postnatural, leveraging the natural environment towards sophisticated forms of artistic intervention.

To treat collaboration in this way is to imagine social activities as an integrated part of artistic practice, a form that can couple and catalyze relationships between science and the imagination. This is evident in the works Willet herself produced during *BioARTCAMP*, a suite of specimens grouped under the heading of *Natural Science* (2011-present). Pieces of hair, leaves, mysterious liquids in test tubes: each given a number as if to suggest that a system

Jennifer Willet & Kira O'Reilly, *Be-Wildering*, 2017. Color photograph, photo credit: Bas de Brouwer.

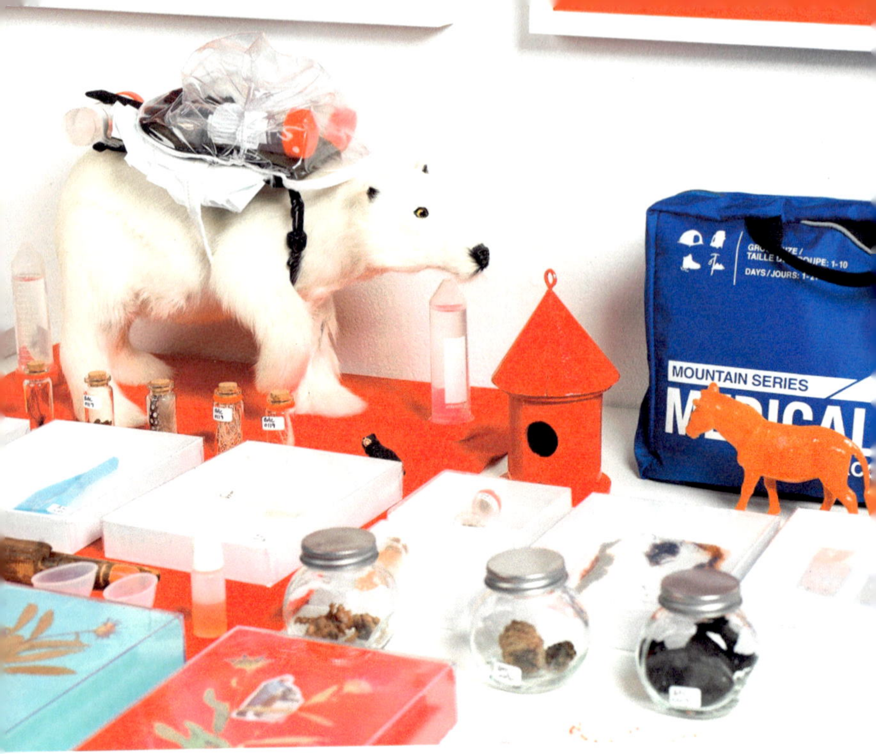
MOUNTAIN SERIES

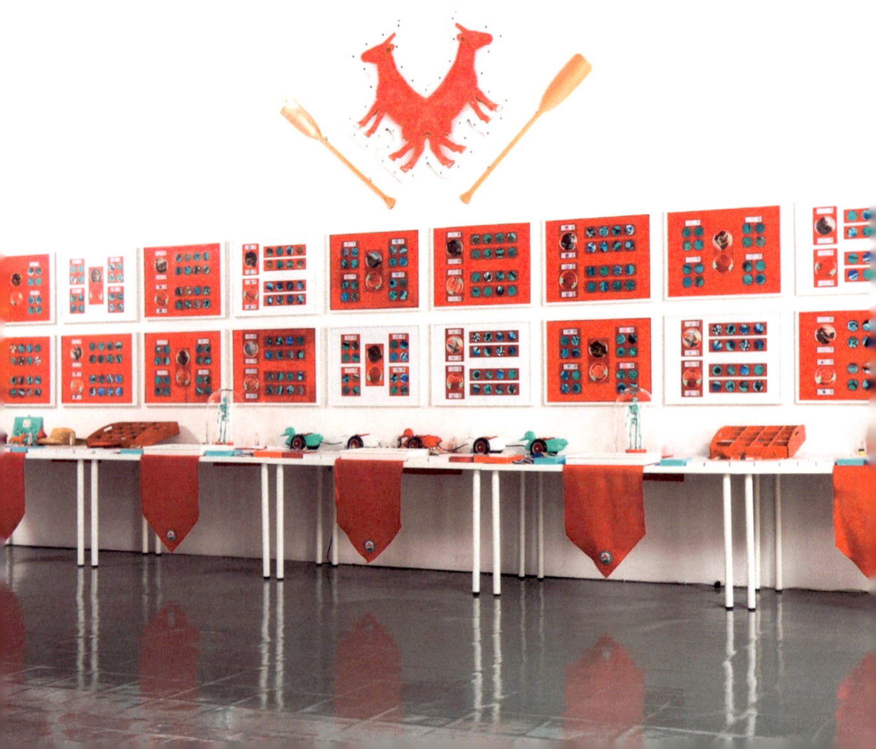

of categorization is in play. Except that no master list is provided and, as such, the linkages between the various specimens in this collection exist less to be classified than to be suggested. The objects wear science as a costume, yet, without the presentation of data to anchor the suggestion of authority, the opposite is also true—and these objects also exist as much to be imagined as to be understood. This is the particular charm of Willet's work, situated as it often is at the intersection of science and the imaginary—insisting that to imagine is also to engage with science as a collaborative partner in the artistic process.

Finally, it is important to suggest that any truly catalytic spirit will always be partly pedagogical in form, attuned to the context in which it interjects itself such as to maximize on the potential for artistic and social impact. For Willet this requires balancing her own creative visions with an institutional imagination that understands bioartistic practice as dependent on protocols of science and principles of biological experimentation. It is significant that she has created the first bioart laboratory in Canada—INCUBATOR Hybrid Laboratory at the Intersection of Art, Science and Ecology—a project that is as much pedagogical as it is creative, situated in the School of Creative Arts at the University of Windsor. Her students engage in certified biological protocols, experimenting with life in artistic ways—extending the theater of artistic engagement to the stage of life itself. Importantly, students learn how to

Jennifer Willet, *Natural Science* (Installation view),
Open Space, Victoria, Canada, 2014.
Photo credit: Arturo Herrera.

perform scientific protocols *and* how to creatively skew (and sometimes contaminate) them, in works of their own and by participating in theatrical events, public demonstrations and parades organized by Willet. This doubled movement—to create and contaminate—is a significant marker of Willet's work seen for its larger trajectory, in ways both metaphoric

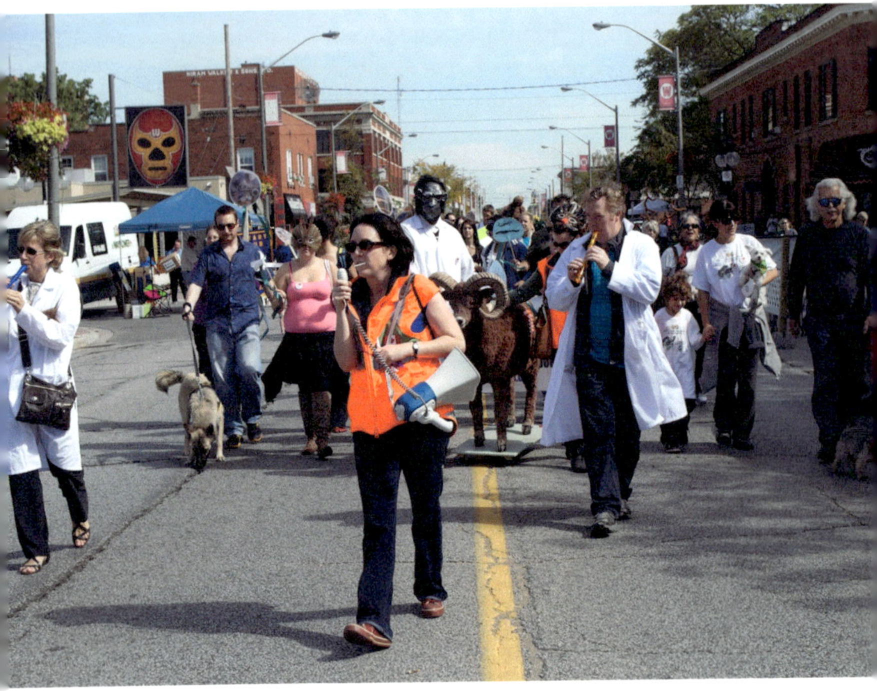

Jennifer Willet, *ECO NUIT PARADE* Documentation. Windsor, Ontario, 2012. Photo: Victoria Symons.

and literal, suggesting a reversibility of terms and a further entanglement of ideological forms. For one of the most interesting things about catalysts is that they become part of the systems they transform—and this is certainly true of Jennifer Willet—working at the edge of the naturalized postnatural.

Naturally Postnatural

Seen through the lens of Jennifer Willet's work, the postnatural is characterized by a sense of tentative and sometimes paradoxical presence: uncertainty is its ontology, performance is its epistemology, and community is its metaphysics. Ontologically, there is no certainty that is not subject to the experimental imagination of scientific possibility, no truth that is impervious to "alternate fact," and no data that cannot be creatively called into question. Epistemologically, we now understand that methods influence outcomes, observation biases results, and the translation of experience is always a valuable companion to the framing of information. And metaphysically, the meaning of our engagements will always receive the warmest welcome from those closest to home, communities that will keep voices vibrant and attuned to one another because they are bound by social proximity rather than obligation. In this spirit of bringing-together, this volume collects a wide array of interventions, expositions and theorizations by a group of artists, scholars and curators bound together in speculative exploration of biology, ecology and artistic practice. Together they represent a complexity of perspectives and a collection of voices brought into proximity by their individual responses to Willet's work and to the question of the naturally postnatural.

The book begins with a provocation by George Gessert and Beth Franks who—in their short story "Vat Room Number 4"—provide a texture for the complexities of the biotechnological question. At the intersection of the technological and the interpersonal, "Vat Room Number 4" is a foreboding tale of a futuristic world in which bodies are preened and harvested in the name of class-based access to technologies of immortality, setting the stage for the challenges of thinking at the intersection of bodies and technological living.

The complex relationship between bodies and power is also paramount to Robert Zwijnenberg's article "BIOPLAY: Clandestine appropriation as an artistic strategy," an extended reflection on an artwork of Willet's in which she performs in a biological fume hood, in the process contaminating (for art) the formerly sterile (scientific) environment. Meditating on the question of "what should we think?" about this work, Zwijnenberg offers the insight that Willet's piece is not about providing answers, but about contamination in a larger sense—and one does a poor job of appreciating contamination if one remains unaffected at the end. Zwijnenberg insists that something clandestine emerges in our encounter with the work, and as a result our role as viewers, participants, and agents of interaction is also to reciprocally contaminate: interpretation itself is a dance of contamination based on the allegory of Willet's *Bioplay*.

Giving different form to a similar ideological impetus, Paul Vanouse's "Deep Woods PCR: Responding to the Challenge of BioARTCAMP" also has something clandestine about it: a DIY insistence on process that is also an acute irreverence towards the sterility normally assumed to be core to the biotechnological project. Recreating complex biological processes over a campfire, Vanouse not only demonstrates a disregard of the scientific pretense towards the necessity of sterility, he also makes good on the promise of a low-tech substitute: a testament to the powers of speculative artistic experimentation. The creation of an alternative procedure may be less the point than the refusal of doctrinaire protocols of thinking and behaving, yet the beauty of "Deep Woods PCR" is, at least in part, the surprise of achieving a functional result.

But surprise can happen in many ways, and disregard too sometimes has consequences—a reminder played out in the first of three contributions to the book by Christian Kuras, "Tuesday 4:41 pm." This short illustrated sequence examines the inner states of a man at work, provocatively gesturing towards the explosive imaginary potential of psychological interaction at the same time as acknowledging the persistence of protocols, whether social or scientific, in everyday life.

An interesting contrast to the idea of protocol is the improvisational contribution by conceptual artists Louise Chance-Baxter& and IAIN BAXTER&—"Untitled (BioARTCAMP)," which consists of a dance through the forest, at times accompanied by a life-size drawing mannequin. In a gesture towards different forms of commune and communication, such actions reinforce the importance of improvisation as a way to construct non-anthropocentric modes of engaging with the world. At the same time, Chance-Baxter& and BAXTER& model ways in which spontaneous actions can retain affective charge, sometime seductive, sometimes playful, but always with a clear acknowledgment of the public site of activity, even when the only people watching aren't people at all. In this sense, "Untitled (BioARTCAMP)" might be thought of as an invitation to engage first, without necessarily needing a clear communicative agenda ahead of time.

The lingering question, of course, is what do we do once we have been given permission to engage? How do we hold ourselves accountable to the new territories of biotechnological engagement opened up in a naturally postnatural world? One answer lies in Warren Cariou's "Petrography: Bitumen in Place" in which Carious reimagines an early 19th century photographic process in order to put the environment into conversation with itself. Cariou creates images of oil, with oil—a process he calls petrography that renders its image by capitalizing on the light-sensitive properties of bitumen petroleum. The results are haunting images of the Canadian oil sands, documenting the landscape "in a more intimate and embodied way, one that resists the distancing effect of spectacle and show us how we as viewers are connected to the damage we are beholding." In Cariou's work, the poetics of technique are brought to the foreground as a guarantee of embodied interaction and accountability.

In a similar spirit, Natasha Myers's essay—"Protocols for an Ungrid-able Ecology: Kinesthetic Attunements for a More-than-Natural History of a Black Oak Savannah"—embraces the challenge to engage with the doubled status of thought and action. For Myers, this is about "doing ecology otherwise," an interesting echo of philosopher Richard Kearney's 1988 challenge to "imagine otherwise" as a response to an overly prescribed culture of information and knowledge.[10] But Myers out-imagines Kearney since key to her framework is not an argument for why such imaginings are needed, but, more radically, the activity of actually doing it. Dancing, listening, recording, photographing—in Myers's vision, these forms of engagement all become embodied practices of living within a participatory ecosystem.

A pair of stories by George Gessert form the next contributions to the volume, reinforcing the complexity of thinking about ourselves as part of a postnatural ecosystem. "Good Housekeeping" and "Hi" are two very different stories that serve as poignant reminders of the reversibility of the biotechnological landscape. As we imagine otherwise, so too are we constantly reimagined by the ecosystems in which we participate. Gessert's stories—accounts of the physical and emotional intensity of living with cancer—remind us that we are also sites within this ecosystem, sometimes sites that are acted upon without our own agency or volition. Taken allegorically, these are not only powerful stories of living in the shadow of invasive disease, they are moments of reciprocity where we are entangled subjects, acting and acted upon by other biological entities.

Melentie Pandilovski's "Phenomenology of (Non) Habitual Spaces for the Bioarts" is, in many ways, an attempt to balance the doubled perspective that emerges between Gessert and Myers—arguing for a nuanced philosophical framework by which forms of lived experience might be differentiated. Asserting that historical tools of analysis are inadequate for understanding new systems of biopolitical power, Pandilovski reflects on the wyas that phenomenology can help anchor experiences that defy familiar modes of apprehension. By focusing his discussion on a selection of bioartists, Pandilovski aims to expand our conceptual possibilities for resisting the productive registers of power and biocapitalism, what he terms the "global biopolitical apparatus." Where phenomenology meets bioart, Pandilovski sees the potential for new modes of perception, catalyzed by technological change and activated by interactive engagement.

In a similar spirit of resistance to historical tools of analysis—particularly *human* tools—Amanda White & Alana Bartol's "Notes from the Deep Earth Treatment Centre" takes embodied experience to a level that is at once literal and speculative. Focusing on soil as a metaphor and as a material entity, White and Bartol seek strategies for communicating with the Earth itself, seen not simply as an abstract concept, but as the literal ground beneath our feet. It is consequently not only human experience that is their focus, but also the possibilities for relating in some way to the experience of soil. Undaunted by the fact that a conversation with dirt may be ultimately impossible, White & Bartol's work suggests speculative and embodied methods for imagining different configurations of the relationship between humans and the (postnatural) world.

A second intervention by Christian Kuras follows: "Ethical Tours & Hunting International," a simulation that is too prescient to really be ironic, but too jaded to be anything else. Promising, for a fee, to make possible the hunting of last members of endangered species, Kuras's provocation serves, at least, to reinforce the ambivalence of naturalized thought in a postnatural age—an age where disaster tourism meets terrorism in the possibilities for extreme vacation. The opposite of a collaborative future, the vision of ETHI is one of postnatural exploitation and the dangers of allowing the human to retain ideological dominance in an era of increasingly liquid biological boundaries.

Reinforcing that these kinds of questions and provocations are anything but fiction, Marta de Menezes's essay—"Dynamic Stasis: A Perspective on Representation in Bioart"—outlines ways in which the difference between audience and actor becomes blurred and ultimately disappears when seen from a bioart perspective. Framed by pieces of her own work—genetically engineered butterflies, living paintings, and sculptures made of neurons—de Menezes argues that the time has come to understand life as already artistic, and art as now having its own capacity for life, dependent on, but also independent from the gaze of the interactive viewer. From André Breton to Nicolas Bourriaud, such a perspective has a robust philosophical and artistic history, but what is different this time is that de Menezes's perpective raises the stakes of the question by literalizing—biologically—the consequences of interaction.

If the complexities and entanglements of these diverse perspectives have any common thread it is not analytic, but poetic, a point best made not by arguing but by actually fusing projects and prose in engaging and thought provoking ways. Such is the focus of Kira O'Reilly's "Unlikely Conspiracies of Biologic Thought," which pairs images of projects and performances—some of the them collaborations with Willet—with anecdotes and narrative vignettes, making clear that the power of engagement in bioart, as in all forms of creative practice, relies on a willing performative spirit. From stories of taxi rides to descriptions of laboratory photographs, O'Reilly's work is a reminder of the autobiographical anchor and the poetic incommensurability of artistic practice, as well as an insistence on the dialogic destiny of art and performance alike. To be posthuman is to animate and be animated; to be postnatural is to understand oneself as a poetic companion to the emergence of new stories, new tragedies, and new possibilities.

This volume ends with a postscript in the form of a final contribution by Christian Kuras, "There is no such thing as gradual change." Change is a common theme among the authors and artists contributing to the book, as well as a site for creative reinvention. To argue against "gradual" change is also to invoke the ways in which change can be disruptive to normative practice, but only when conceived as such. Gradual change will always run the risk of being subsumed by the institutional rhetoric of "progress." A far preferable future—postnatural though it may be—is one absent from such rhetoric, where change is anything but gradual, often surprising, and sometimes disconcerting. When "there is no such thing as gradual change," we are left—to the best of our collective abilities—to naturalize for ourselves the postnatural possibilities of contemporary living.

This is a book about life in a state of catalysis—a recognition of the entangled complexities of engagement in a time when the (environmental) stakes and the (technological) possibilities of living have never been greater. The challenge of this book is to treat the question as one that is both "live" (in the sense of immediate, pressing and unreconciled) and "living" (in the sense that real, material, entities—including us—are implicated in varying registers of engagement). This is the moment where aspirational philosophies such as Jane Bennet's "vibrant materialism"—an eloquent invitation to treat all matter as vital[11]—meets the daily pragmatics of biological living, including new possibilities in genetics and their sometimes unexpected mutations, manifestations, and what Richard Pell and Lauren Allen call "push back":

> Our environment is constantly sculpting us; the changes we make to organisms have consequences

> for how humans conduct themselves. In nearly every case, the changes humans have made to an organism push back against us and inspire further changes to the constitution of the human.[12]

This is to say that there is something wonderfully tangled about the postnatural, situated between the (anthropocentric) end of the world and new (genetic) possibilities for engineered life. Pushing Beuys's prophecy one step further, no longer are we simply creating social order as an artwork of the future. This future has arrived and we ourselves have become living parts of the artwork that is our collective present, postnaturally naturalized in the collaborative and collective act of creative production.

Notes

1 Julie Hannon, "Earth in the Age of Humans," *Carnegie Magazine*, December 2016. http://www.postnatural.org/Earth-in-the-Age-of-Humans

2 Donna Haraway argues something similar to this in her book, *Staying with the Trouble*. Because of the environmental and political charge the term "Anthropocene" has accumulated, Haraway proposes an alternate: the Chthulucene. I am less interested in the specifics of her alternate proposal than the fact that she makes one—troubling the opacity of scientific terminologies in the name of an alternate fiction. This is significant because to counter "truth" with "fiction" is a decidedly posthuman and postnatural strategy—an ideological shift away from the logic of analytic accountability and towards the making of "kin" (Haraway) or more simply, community. See Donna

Haraway, *Staying with the Trouble: Making Kin in the Chthulucene* (Durhan: Duke University Press, 2016).

3 Zara Stone, "Everything You Need To Know About Sophia, The World's First Robot Citizen," *Forbes*, November 7, 2017. https://www.forbes.com/sites/zarastone/2017/11/07/everything-you-need-to-know-about-sophia-the-worlds-first-robot-citizen/#8922ec246fa1

4 Harry Pettit, "Scientists implant tiny human brains inside rats to give them 'enhanced intelligence' provoking an ethical outcry," *Dailymail.com*, November 8, 2017. http://www.dailymail.co.uk/sciencetech/article-5061891/Outcry-scientists-implant-tiny-human-brains-inside-rats.html

5 Laurie Rojas, "Beuys' Concept of Social Sculpture and Relational Art Practices Today," *Chicago Art Magazine*, November 29, 2010.

6 I make this argument in more detail in "Ecologies of (Imaginary Friendship)" in Ted Hiebert, ed. *Plastic Blue Marble—Catalyst: Amanda Boetzkes* (Seattle & Victoria: Noxious Sector Press, 2016), 79-96.

7 While BIOTEKNICA started with digitally simulated teratomas, they did go on to grow living sculptures, most specifically a mouse teratoma cell line grown on a bio-adsorbable polymer sculpture of a teratoma (in collaboration with Oron Catts, Ionat Zurr, 2006).

8 BioARTCAMP participants included: Iain Baxter&, Angus Leech, Tagny Duff, Paul Vanouse, Marta De Menezes, Marie Pier Boucher, Kurt Illerbrun, Bulent Mutus, Jeanette Groenendaal, Zoot Derks, Jennifer Willet, Jamie Ferguson, Britt Wray, Kacie Auffret, David Dowhaniuk.

9. Jennifer Willet, "BioARTCAMP Project Statement," 2011. http://jenniferwillet.com/home/projects/bioartcamp/

10 Richard Kearney, *The Wake of the Imagination: Toward a Postmodern Culture* (London: Routledge, 1988).

11 Jane Bennet, *Vibrant Matter: A Political Ecology of Things* (Durham: Duke University Press, 2010), vii.

12 Richard W. Pell & Lauren B. Allen, "Preface to a Genealogy of the Postnatural," in Anna-Sophia Springer & Etienne

Turpin, eds. *Intercalations 2: Land & Animal & Non-Animal* (Berlin: K. Verlag and the Haus der Kulturen der Welt, 2015). http://www.postnatural.org/Preface-to-a-Genealogy-of-the-Postnatural

1 Vat Room Number 4

George Gessert & Beth Franks

After the election my thoughts turned to food. I treated myself to childhood favorites, waffles, strawberry sundaes, cream of wheat with brown sugar and half-and-half. But whatever I was seeking was not there so I tried poppy seed pastries and tiramisu. Then came scalloped potatoes baked in heavy cream, coq au vin, eggs with béarnaise sauce, frozen lychees, shrimp bisque and bagels piled high with smoked trout. I ate out often, sampling everything from Ethiopian, Szechwanese and Peruvian, to BBQ at Porky's Palace. Certain restaurant meals stand out, like the time at Xenon when I had Tiger Tears, a salad of fresh greens with julienned vegetables in a lime-chili dressing, topped with paper-thin slices of beef so rare they were almost raw. The main course, halibut in a macadamia crust, was excellent but I would have quickly forgotten it if not for the Reisling. It struck sugary, discordant notes far more apropos than harmony. I had become fond of sweet wines: ice wines, spumantes and especially muscats, but as November days grew shorter and winter torpor settled in, I needed caffeine. I no longer limited myself to two espressos a day, but sipped them all day long. Occasionally I'd slip over to the Beanery for a Buzzilla, a blend of whole milk, vanilla Haagen-Dazs and three shots of espresso, topped with whipped cream. Then I'd read the *New York Times* and revert to a childhood habit of chewing my nails. I'd mince the fragments between my front teeth into a grainy, gelatin-flavored mash.

What troubled me wasn't so much the president, appalling as he was, as the people who had voted for him. What were they thinking? What did they want? Did they imagine that a cartoon could deliver them from reality? Lost! And myself, how many of my own hopes and dreams were illusions? Consciousness had become a guerrilla exercise full of traps, blind alleys and ambushes, so I turned to pears and St. Andre cheese.

Everything mutates, even food preoccupations. My mutagen was an installation: *Bioteknica: Organic Tissue Prototypes* by Jennifer Willet and Jason Knight. The work mimicked corporate advertising and annual reports, and drew me into a gleaming world of medical research. Sterility has powerful aesthetic appeal, but as I perused glossy photographs of laboratory glass and stainless steel, it dawned on me that the golden liquid in one of the test tubes might be urine. Nearby, a lump of gristle sprouted hair, and suddenly I remembered vat room number four.

I was 13 and an avid science fiction fan. *Synthetic Men of Mars* by Edgar Rice Burroughs is about a megalomaniacal warlord who builds an army of warriors gestated in vats. Only about 50% of the "hormads" are functional—the rest are killed—but 50% is enough to fill the ranks, so the project goes full speed ahead. Then something goes wrong, terribly wrong, in vat room number four.

The mutagen worked slowly. Fully six weeks passed before I sought out the yellowed book (published in 1939) in the public library. The narrator enters the vat room and discovers that "instead of individual hormads being formed, there was a single huge mass of animal tissue emerging from the vat and rolling out over the floor. Various internal and external human parts and organs grew out of it without any relation to other parts, a leg here, a hand there, a head somewhere else."

The narrator orders the room sealed, but as he leaves he sees one of the heads take "a large bite from an adjacent piece of tissue ... it wouldn't starve to death." Soon the mass bursts a window and spills out into a courtyard, threatening to bubble into surrounding swamps, and from there to overrun Mars.

***

I wanted to share my find with Beth, my sister, who had introduced me to Edgar Rice Burroughs when we were in junior high school, but as I began an email to her, something strange happened. One by one and then in little clusters, words appeared on my computer screen as if by themselves. My fingers danced over the keys, possessed by a mysterious force, something outside me, something alien. In retrospect I'm pretty sure that this alien force was my discombobulated digestive system, but at the time such a possibility did not occur to me. I was aware only that as words emerged in the field of light, they transported me into the future.

The alphabet is privatized letter by letter, along with genetic information, species, sound waves and the spectrum, visible and invisible. Oligarchs engage in internecine corporate struggles and display their wealth in fabulous feasts.

The lives of commoners revolve around family, food, ceremonial shopping, fears of aging and alien invasion, and observance of taboos, such as never using the word "death," and never talking with your mouth full. Most commoners despise the less fortunate, who are called "the faithless ones." When distractions fail, slogans kick in, like "Shopping makes us one with God." Or is it Goddess? Coatlicue?

Notification comes by mail, old-fashioned for gravitas. Only those in the prime of youth are called. They report to induction centers, and after thorough physical and psychological examinations the Select are assigned to special resorts.

Model for the resorts: pastel-colored time-share condominiums north of Puerta Vallerta built in the style of Aztec pyramids. Manicured lawns, swimming pools, cabanas, drinks with little parasols (do they have little parasols in the future? Maybe twinkling swizzle sticks?), and white sand beaches. Here the Select lounge and play.

The serious work is eating. Every meal is compulsory. Whoever misses a meal must report for a transfusion. The food is delicious, drenched in oils, and laced with additives, many of them classified secrets. A few months of this diet produces marbled, pre-seasoned meat. It is hallucinogenic, like the flesh of reindeer that have eaten amanitas. The meat also contains anti-aging factors. Oligarchs consume enough of this meat to extend their lives, which are on average several times longer than commoners. On very special occasions a few commoners are rewarded with a taste.

The Select, like all commoners, are conditioned from birth to make light of it all. On the morning of her dispatching a young woman might devote her final hour to deciding whether to wear her memory lane musical skirt, or the one decorated with kinetic teddy bears. Seriousness is acceptable, within limits. The same young woman might admit that she isn't sure if she believes in transfiguration, but she does believe in her family and the Greater Good. Those who resist are considered cowardly, parasitic, depressing, and unclean, in a word, evil. They are fed to crocodiles.

After I emailed this to Beth, I ate at Jade Pavilion and ordered my usual, Ants Climbing Tree Special, plus Honey Walnut Shrimp with melon and sweetened mayonnaise. The shrimp came with lightly steamed broccoli, which I set aside. Good nutrition was not the point.

A few days later my sister emailed me a vignette about Dug, a young man who has received notification to report.

There's been a party every night since I received the letter. Messages pour in, and phone calls—not just postings—from people I haven't heard from since high school. Still, I must confess that sometimes I feel a little anxious.

The evening before I reported to the induction center, Dad took me aside. "We're so proud of you, son," he said. "Your mother and I ...." He couldn't finish, but turned away, cleared his throat and coughed into a paper towel.

Mom was at the macrowave. She wiped her hands on her apron and said, "The priest mentioned you in Wednesday Circle." She touched my shoulder. "Just look at you," she said. "My tall, handsome son."

That night I took a long look at myself in the mirror. I have to admit that I do look good. I work out every day. Coach set up the program for me and it's succeeded. I vowed then and there that I would make them proud.

***

Dug reports to the induction center, passes with flying colors, and is sent to a resort. A few days later he writes his parents:

"After breakfast, which is good, believe you me, comes marathon shopping. Every day we're driven to a different mall. Each of us has a target expenditure, and has to buy everything before lunch. You know some of this already, but it bears repeating: if I don't spend the full amount, nothing gets shipped out. Nada. And I'm not allowed to duplicate. There's a lot of planning involved, like getting sizes and colors exactly right—no returns! Mom and Dad, you've done such a great job! Keep those lists rolling in."

Dug buys a golf cart with a built-in aortic pulse monitor for his father, and a cashmere sweater and Babycakes Rinkel Therapy for his mother, not the lotion but the implants, top-of-the-line. For little cousin Visa he gets a four foot robotic panda with peek-a-boo TV eyes.

***

Over the next few weeks Beth and I refined the story. We decided, or perhaps I should say realized, that when Dug admires himself in the mirror, he feels wistful about his hair. He knows it will soon fall out. Hormones in the diet will pre-pluck.

At the induction center, Dug momentarily reflects on his life. He tells himself it really is like the posters say: Only the Best and Brightest. But he knows this isn't true—not in his case, anyway. Except for gym, he never did well in school. He broods about the time in the third grade when he stole some chocolates from Exxon Tetra, a brainy rich kid who sat in front of him. The teacher found the chocolates in Dug's desk. That was the first humiliation, proof of his stupidity: he hadn't thought to hide the candies more carefully. The entire class listened as the teacher questioned him about how they got into his desk. He insisted that he didn't know, then burst into tears. Exxon cut in, "Maybe they washed in on your tears?" After that no self-defense seemed possible.

***

Dug is swept up in a round of feasts, shopping, and partying. He writes home:

"Sunday Brunch is tops: Flank steaks, real ones, not vat-grown, lobster double dip fondue, potatoes swimming in actual cow butter and sour cream, and piles of desserts. The caramel fudge pie and marshmallow coconut cheesecake are my favorites. Next week they're introducing a line of ultraplus foods, top-secret stuff still in development. You won't find these in stores, not anytime soon, but they said I can tell you. Keep this strictly to yourselves: Kouch Potato Pancakes with steroid butter and tingle jelly foam, deep fried duck glazed in special hormones that give you the blind munchies, and BBQ pork drenched in Bunkerbuster Sauce. Yes, Bunkerbuster!"

Dug mentions Chucky, a new friend. Chucky has been at the resort for almost two months and is losing his hair—his date is set—and loves to party. Dug relates some of Chucky's antics, but doesn't mention that one evening Chucky stole some BBQ from the dining hall. He'd been drinking, as had Dug. After Chucky passed out, Dug dipped his finger in Bunkerbuster and took a taste. It was so delicious he couldn't stop eating until it was all gone. In the process, Dug slopped some sauce down his shirt. The stain wouldn't wash out, so he buried the shirt beneath a frangipani. No one could find out, not even Chucky, because eating Bunkerbuster before receiving clearance is strictly forbidden. Flavors might not develop properly. Chucky had warned him: once you've tasted Bunkerbuster you're addicted. Dug discovered for himself that it was true. He couldn't stop thinking about the sauce.

Meals begin with prayers, which drone on without pause through dessert. So great is the honor of hearing these sacred incantations, delivered only during The Fattening, that few among The Select are not profoundly moved even if they don't understand a word. Wall-sized screens surround The Select with images of paradise, which is an alabaster shopping arcade extending to infinity in all directions.

The Select become mystic brides and bridegrooms of the corn. Dug's roommate is a young woman his own age. The two were assigned for their spiritual, emotional, and sexual compatibility. The induction center, with its long experience and comprehensive physical and psychological profiles, rarely makes a mistake. The system serves two functions: happy pairs tend to eat heartily and fatten on schedule. And stories of perfect matches leak out to stoke teenage fantasies of joining the Select.

***

Beth and I agreed that the dispatching is a sensitive detail. A jungle hunt would not work because it would recall Richard Connell's "The Most Dangerous Game." Also there won't be any jungles left. But most important, who wants to pick buckshot out of a hallucinogenic steak? Game, some gourmets say, can be subtly flavored by adrenaline and fear. The best meat (the Aztecs favored hearts and hands) is from those who have been peacefully slaughtered. No pinpricks of anxiety, not a trace of pain. Hymns, soft lights, vintage wines infused with sedatives.

***

The roommate is from a priestly family fallen on bad times. One day she tells him that she sometimes fantasizes about escaping. He is torn. Should he tell the crocodile police?

Late that evening, as they lounge together in bed and share a snack of coconut creme corndogs with maraschinos, she talks with her mouth full. He gets only a glimpse of masticated corndog, but that is all it takes to reveal the horrifying truth: she is evil. Irredeemable. Maybe she has been playing him all along.

"I'm leaving tonight," she says. After a long pause she adds, "Come with me."

A spot of whipped cream clings to her upper lip. Dug conceals his revulsion. "I can't," he says, shaking his head. "It would make everything ... meaningless." She tries to convince him, but he refuses to listen.

"So you're going to report me?" she asks.

"No," he says. "I'd never do that." Which at that moment is true. "But don't leave. You can't." She gathers a few things, then glances at him and seems to hesitate. She darts over and kisses him.

"Love you," she says, and is gone. Dug waits for a full twenty minutes (he checks the clock) to give her time to change her mind. Then he phones the desk.

"We're on it," the matron says. "You're down for extra Bunkerbuster."

He feels sick. He hopes that this is a test. Tomorrow she'll be back and they'll tell him that he's passed. "But what a bitch," he sobs, "to put me through all this!" In a fit of helpless rage he tears his hair. It comes out in clumps.

The story ends as a train moves through mountains. A gaunt, blackened figure sits in a coal car, gazing out. Clouds rest on peaks, but otherwise the day is gloriously sunny. A flock of birds passes overhead. She looks up and smiles.

2 BIOPLAY

Clandestine appropriation as an artistic strategy

Robert Zwijnenberg

In spring 2008 Jennifer Willet taught a bioart class I had organised at Leiden University entitled Contemporary Art and the Life Sciences.[1] During her stay in Leiden, Willet did a performance at a lab in the Gorlaeus Laboratories, in collaboration with the Arts and Genomics Centre.[2] Besides Willet, the only other person at the performance was American bioartist Adam Zaretsky, who was there to take pictures of the event. One of Zaretsky's photos of the event hangs in my home, nicely framed. It shows Willet lying undressed in a fume hood. The picture elicits a range of responses. Guests with knowledge of art history see it as part of the tradition of the nude in Western art. The photograph's aesthetic qualities certainly justify such an interpretation. To most guests, however, it is unclear where the picture was taken, or what a fume hood is, let alone why anyone would get inside one. The responses tend to be humorous—apparently a naked woman in a fume hood can be funny. But what do we actually see in the photo, and how can we interpret its meaning?

Willet describes in an article how the performance, entitled *Bioplay*, came about. "*Bioplay* ... consists of a series of digital photographs which document clandestine performances in laboratory environments where the artist removes her clothing and conducts biotechnological protocols in the nude."[3] Willet took off her clothes and climbed into the fume hood in the lab, where Zaretsky photographed her. For those who do not know, a fume hood is an enclosure with adjustable access for the user's hands and controlled air circulation that enables a researcher to work with hazardous substances. It protects

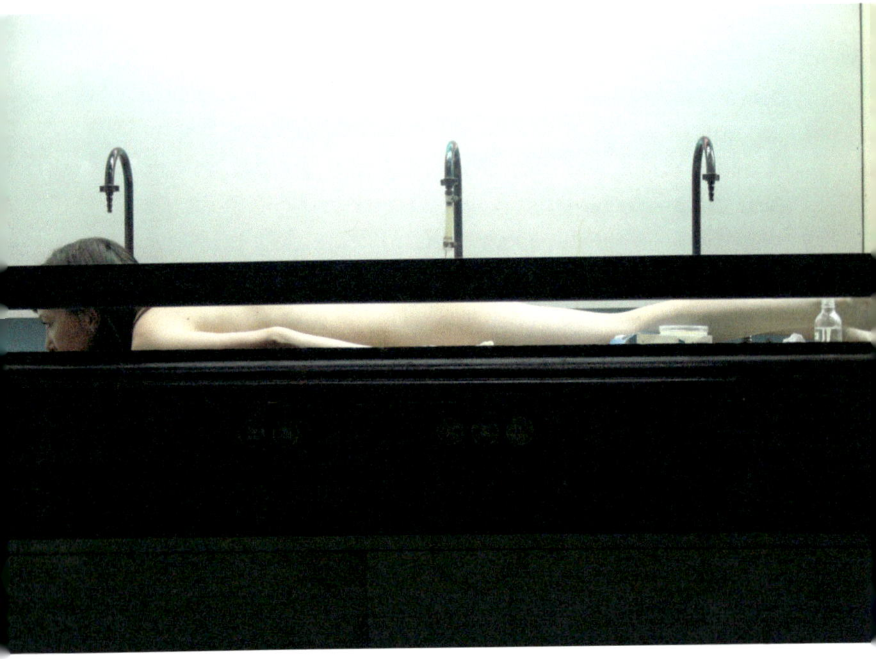

Jennifer Willet, *Bioplay*, 2017.
Color photograph, photo credit: Adam Zaretsky.

the researcher and it ensures that the materials being studied are not contaminated by the researcher's body. The "biotechnological protocols" to which Willet refers consisted of taking samples from various places on her body using a cotton swab, then allowing the bacteria from the body to grow in gelatin nutrient agar plates for a week.

A first question that might be asked is: what is the artwork here? In her article, Willet describes her activities in the lab as a "body of work."[4] This describes the nature of the work, but also invites us to consider what body or kinds of bodies there actually are in the work. There is the artist's body at work in the fume hood and there is the photographer's body at work documenting. Willet complains in the article that the fume hood was dirty and that it took her an hour to clean it. She does not therefore simply surrender to the place, but appropriates it by adapting the hood to her own personal standards of hygiene. It is also significant that the photographer is Adam Zaretsky, one of the most important bioartists of our time, and certainly also one of the most controversial.[5] He witnesses the performance, but he also translates it into images: the represented body at work in the pictures. These are clearly staged photographs, the artist adopting poses that give the pictures an aesthetic quality that is enhanced by the position of the camera and the lighting.

A third body is also involved in this work: the body of the artist as writer. I think it is not really possible to ignore Willet's article about her clandestine performances as an aspect of the work itself. In the article she gives a precise and detailed description of her intentions and goals for this work. The article therefore reads like an artistic manifesto in which she seems to be setting out the direction of future work. Willet describes her intention with the performance as "propagating alternative models of biotechnology towards a wider representation of possible practitioners (artists, mothers, accountants and swimmers) and a wider range of possible relations between the various orders of life that make up the laboratory ecology."[6] An unclothed body posed in a fume hood certainly gives an alternative representation of laboratory ecology, and for Willet this is necessary—as she says in the article—in order to link biotechnology, life and culture: the cultural embedding of biotechnology, in other words.

The works she produced subsequent to *Bioplay* all appear, I would argue, to revolve around the appropriation of what Willet refers to as biotechnological ecologies. Her intention is to magnify, widen, and transform the relations at work in these ecologies into relationships with life beyond biotechnology. In her work, Willet hopes to break through the isolation of the biotechnological from other forms of ecology. In a certain sense, the empowering nature of her art lies in this attempt to embed the biotechnological within the everyday ecologies in which we all already exist. This happens as a result of her appropriation of the biotechnological in order to explore and interrogate it within artistic and social domains, to take control of it, reveal its complexity and ambiguity and expose the self-imposed isolation of the biotechnological. The title *Bioplay* underlines the playful and humorous element that Willet introduces into her work, playfulness and humor that contrast with and question the efficiency and seriousness of laboratory practice.

What exactly is the relationship between the performance, the article, and the photographs? The aesthetic quality of the photograph may, as I have said, lead some to place it in the tradition of the nude in Western art, but this does not do justice to Willet's apparent intentions. Her article gives what at first glance seems to be an exhaustive account of these intentions. It can be read as a theoretical basis for and legitimization of the performance. The question is whether the text obscures the meaning and relevance of the performance as a necessary and factual event? In other words, why was the performance necessary? What does the actual performance add to the arguments and conclusions in Willet's article? In asking these questions I am looking for the value and meaning of Willet's actual bodily investment in the work. Could one approach *Bioplay* in such a way that the actual performance, rather than the article and the photographs, grounds the artistic urgency and the theoretical relevance of the project in the cultural and societal debate on biotechnology?

For me, the urgency and relevance of the performance lie in what Willet describes as the clandestine nature of the physical appropriation of a biotechnological practice. The clandestine aspect of the performance lies in the fact that her activities ignore all health and safety regulations applicable in such a lab, such as wearing a lab coat and closed-toe shoes and the observance of a certain degree of decorum and professional conduct. To lie naked in a fume hood is in fact to ridicule decorum and professional conduct. It is inappropriate behavior that can only take place clandestinely, and is itself clandestine. The clandestine element also lies in the simple fact that Willet had no permission for these performances—and would probably not have received it had she requested it. This may appear a trivial observation, but to me the question of the meaning, the urgency and the relevance of this work are immediately associated with this issue of clandestine access to and appropriation of the laboratory environment, in the sense of something done in secret, to conceal an illicit or improper purpose. I argue that the clandestine nature of her bodily performance gives an extra dimension to the way Willet provides an alternative representation of biotechnological practice.

To look at *Bioplay* from the perspective of this clandestine physical appropriation is to also provide an interpretive lens for much of Willet's later work. A work like *An Incubator in Sheep's Clothing* gains in artistic depth if it is viewed from the perspective of clandestine appropriation.[7] At first, this work appears to be based on the inside-outside principle—the *Umwertung aller Werte*—of biotechnology, another recurrent principle in her work. The concept of clandestine appropriation defines the work more sharply as disruptive, as I shall discuss later in this article. In both *An Incubator in Sheep's Clothing* and *Bioplay*, an artistic reality is grounded in an existing reality of biotechnology. This artistic reality refutes and criticizes biotechnological practice.

Why is the clandestine element so important to this interpretation? Clandestine physical appropriation as an artistic strategy responds firstly at a conceptual level to questions concerning art that engages with the implications of biotechnological practice. In *Bioart and the Vitality of Media* Robert Mitchell distinguishes between two modes or tactics that an artist can employ in the attempt to engage with the life sciences: the prophylactic and the vitalist.[8] In the prophylactic tactic artists use "non-

Jennifer Willet, *An Incubator in Sheep's Clothing*, 2011. Sheep sculpture, incubator, live yeast cultures.

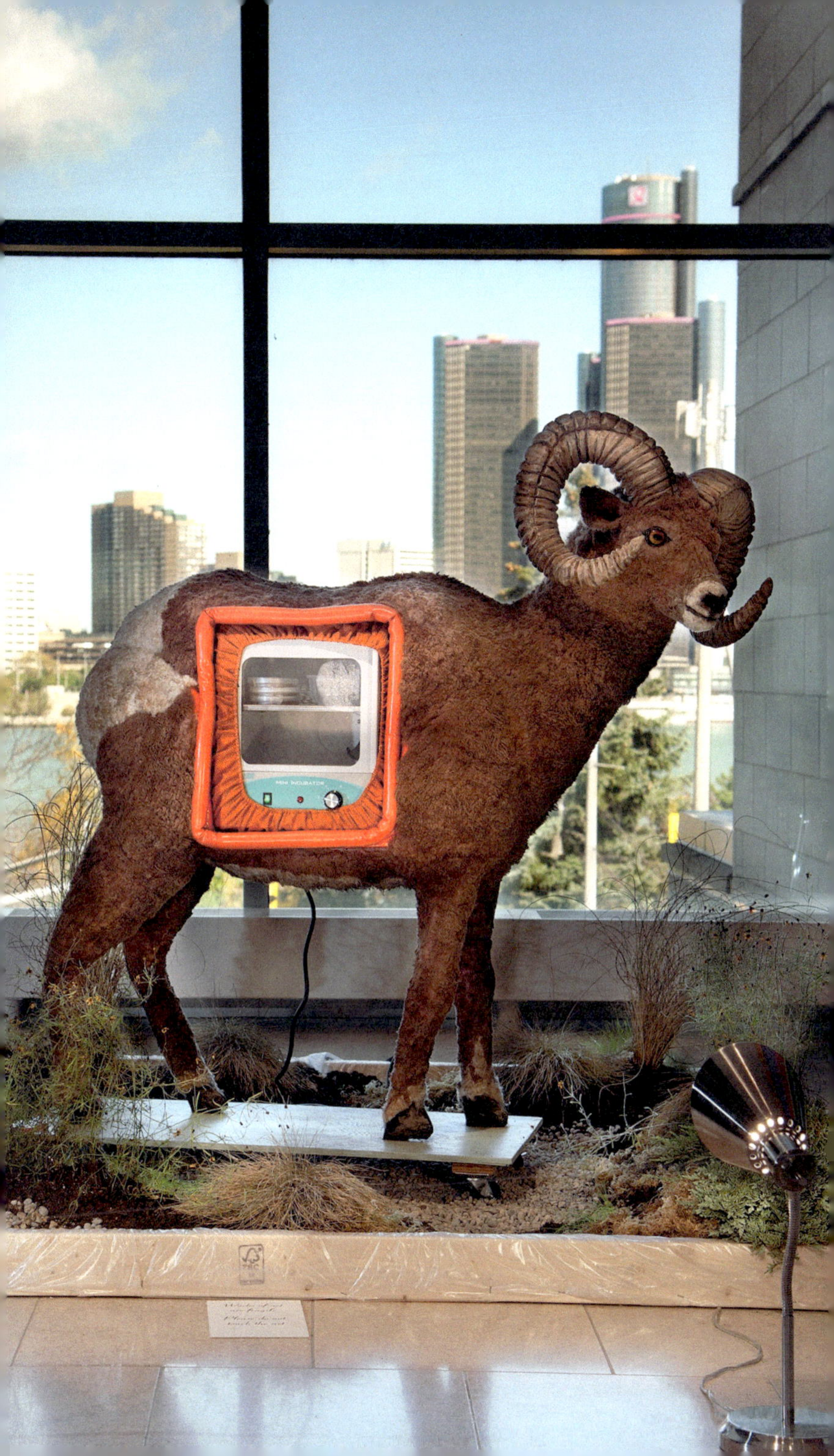
MINI INCUBATOR

biotechnological media, such as paint, sculpted wood or metal, and photography, in ways that allow them to re-present aspects of biotechnology."[9] By contrast, "the vitalist tactic is premised on the principle that art best engages the problematic of biotechnology when it becomes itself a medium for this latter."[10] In Mitchell's definition vitalist bioart is art produced in the laboratory using biotechnical materials and tools. Although the prophylactic tactic also raises all kinds of ethical questions, Mitchell suggests that the hands-on use of biotechnological media leads more directly to the critical exposure of ethical issues. Vitalist bioart is therefore often regarded (by bioartists and critics) as a critical movement whereby artists distance themselves from what happens in the life sciences. However, this artisticocritical position is based on the grounds that these artists produce art in a biology lab, with living materials, and are entirely subject to all the requirements, procedures and licences that also apply to the scientists working in that same lab. This creates a paradox: how can artists establish critical distance when they are subject to the same rules of the lab, essentially unable to do anything that life scientists themselves cannot also do?

The ways that bioartists manage to circumvent or defuse the paradox Mitchell highlights provide indications of how they are truly able to build critical and alternative artistic realities, using biotechnological practice as a basis. This is an urgent issue since for several years the life sciences have been running projects, awards and competitions in which artists are invited to collaborate with life scientists. The life sciences have thus discovered bioart as a means of enhancing their own public image.[11] On the website of *The Bio Art & Design Award*, for example, it says, "Artists and designers interested in the life sciences are invited to propose new projects for funding ... to fully realize a new work of art or design that pushes the boundaries of research application and creative expression."[12] In this competition, Mitchell's paradox is resolved, as it were, as art serves to explore "exciting new intersections among design, artistic practice and the life sciences" and to push "the boundaries of research application and creative expression" rather than seeking that (paradoxical) critical distance. Furthermore, increasingly collaboration between artists and scientists takes the form of co-creation and co-design in response to public disquiet about the ethical and cultural impact of technological innovations. By involving all stakeholders (science, industry, policymakers, citizens) in the development and implementation of technological innovations at an early stage, it is believed that public acceptance will be easier to achieve because safety concerns and people's ethical reluctance have been addressed in the design process. These concepts are surrounded by an air of optimism: if we simply work together we can solve all problems. Artists are also invited or challenged to play a role in these co-creation or co-design processes. In fact, *The Bio Art & Design Award* is a call for co-design.

Whether this is a bad thing or not depends on the answer to another question: what is the role and function of art in today's society? Must art always be urgent and relevant in public and academic debates on the implications of technological innovations? In the view whereby art is a participant in the processes of co-creation, art seems to be regarded as part of a shared social discourse. Though different views are possible within the process of co-creation, the collaboration implicitly assumes a shared background of values, norms and attitudes that enable the process. At the same time, from a romantic point of view, art is held to be capable of offering a new and unexpected perspective on technological developments, due to the fact that it adopts a certain social distance, "to push the boundaries of research application and creative expression."

However, the role and function of art vis-à-vis biotechnological developments can also be explored on a more theoretical level. One author who has addressed this question is Krysztof Ziarek, who says that "in the context of this thinning boundary [between art and technology], it seems legitimate and necessary to ask whether and to what extent transgenic art is complicit with the manipulative flows of power or whether, on the contrary, it exposes, complicates, or perhaps even contests them."[13] Ziarek describes art as a force field, where forces drawn from historical reality and social reality come to be shaped into an alternative set of relations. It "constitutes the force of art [in] its specific capacity for reworking the categorical determinations of reality into a transformative event." For Ziarek, art must play a transformative and critical role in public debate without becoming an accepted part of that debate.

Joanna Zylinska identifies the same paradox as Mitchell:

> Although the two [bioart and science projects] are often developed from within the same labs and are part of the same research grants, bioart's mission is ostensibly different from the one embraced by the biotechnological industry. The primary business of bioart is the representation, articulation and open-ended creation of new forms and modes of life—not capital-induced production of Life™.[14]

The conclusion she draws from this paradox goes beyond Mitchell's concerns about critical distance: "In remoulding life, bioartists cannot therefore be judged by the established normative criteria" but take responsibility for life "without retreating to any predefined entrenched moralist positions about what this life is and how it should be treated."[15] She adds: "I want to suggest that, far from being immoral or irresponsible, artists experimenting with life are themselves performatively engaged not just in enacting life differently but also in enacting a different ethics of life." Taken together, Ziarek's and Zylinska's views rule out the possibility that art can be part of a process of co-creation. Bioart has to account for the paradox that Mitchell identifies, and cannot come to a conclusion other than an insistence that art must actively resist the dominant consensus and be a critical and reflective transformative force.

Chantal Mouffe agrees, elaborating on the question by asking: "Can artistic practices still play a critical role in a society where the difference between art and advertising have become blurred and where artists and cultural workers have become a necessary part of capitalist production?"[16] Mouffe proposes an agonistic approach where:

> The objective is to unveil all that is repressed by the dominant consensus ... to envisage the relation between artistic practices and their public in a very different way than those whose objective is the creation of consensus, even if this consensus is seen as a critical one. According to the agonistic approach, critical art is art that foments dissensus, that makes visible what the dominant consensus tends to obscure and obliterate.[17]

To refer to these thinkers so briefly does not, of course, do justice to the theoretical depth and richness of their ideas, nor the differences among their views. But what they have in common is a focus on the possibility that a theoretical *Umfeld*, or setting, exists from which Willet's practices can be viewed, and which can in turn be enriched by her artistic practice.

By physically gaining access in a clandestine way to the space where biotechnological practice takes place and by appropriating it in an equally clandestine performance, Willet circumvents Mitchell's paradox. She creates her own reality, her own invented performative realm, overlayed upon the biotechnological reality of the lab itself. Installing one's own (artistic) reality overtop of an existing (social or contextual) reality assumes a "primacy of action" or a "primacy of doing" as an essential prerequisite. This absolutely renders the clandestine performative action of *Bioplay* urgent and relevant for the project of creating artistic meaning. The photographs of the performance are the visual evidence of the clandestine activity; the article is the written confession. Together, they provide overwhelming evidence that the performance took place as a wilful clandestine and incontrovertibly necessary intervention.

Willet's work clearly uses both prophylactic and vitalist tactics, working sometimes with biotechnological media and other times with more traditional artistic media. But if we regard clandestine appropriation as Willet's foremost artistic strategy, we can see the continuity between her different artistic engagements. In other words, in Willet's work clandestine appropriation is not merely an actual physical activity, it also guides her work in conceptual terms. In *Windsor Yeast* (2008-2016), for example, she grows microbial samples in Petri dishes. Collected from the post-industrial urban landscape, the smell of the samples "conjures up memories of fermenting wine and beer with her father, in the basement of her childhood home."[18] Growing microbial samples in Petri dishes is usurped by a personal reality of a poetic kind that disrupts the instrumentality of this particular biotechnological practice.[19] Equally, *An Incubator in Sheep's Clothing* (in which Willet builds an incubator into a sheep) appears on first consideration to be a merely anecdotal and humorous act. However, this work gains artistic depth when considered alongside *Bioplay* and *Windsor Yeast*. It is a prophylactic version of a vitalist tactic of clandestine appropriation. The artwork is at once a clandestine appropriation of the sheep's body by biotechnology—biotechnology presenting itself in the guise of an innocent sheep—and a clandestine appropriation of biotechnology by the sheep (its simplicity ridicules the seriousness of biotechnology). Either way, the ambiguity, whimsicality, and humor in the work is the foundation for an artistic reality that reveals biotechnological reality as the sometimes gruesome exploitation of life.

So, to end where we began would be to ask what I should say to my guests who are surprised or irritated when they see the photograph of Willet's *Bioplay*? Given the arguments I have presented here, I should probably say that anyone who looks at art must do so with an embodied and conceptual clandestine appropriation of the artwork in order to discern a meaning that is of existential significance to their own life. In essence, a viewer should play in the fume hood of the image itself.[20]

Notes

1 Contemporary Art and the Life Sciences, course website: http://leidenbioart.blogspot.nl.
2 Art and Genomics Centre website: http://www.artsgenomics.org.
3 Jennifer Willet, "Bioplay: Bacteria Cultures," in Andy Miah, ed., *Human Futures: Art in an Age of Uncertainty* (Liverpool: Liverpool University Press, 2009), 20-23.
4 Ibid., 21.
5 For more on Adam Zatetsky, including samples of his work, see: http://www.emutagen.com. For a video of his performances in the Gorlaeus Lab, see: https://www.youtube.com/watch?v=mve5b8RW6_8.
6 Willet, 21.
7 *An Incubator in Sheep's Clothing* is one of the works from the *BioARTCAMP* and *NATURAL SCIENCE* projects, exhibited in the 2011 AGW Biennial at The Art Gallery of Windsor (Windsor, Canada, 2011).
8 Robert Mitchell, *Bioart and the Vitality of Media* (Seattle and London: University of Washington Press 2010), 26-34.
9 Mitchell, 27.
10 Ibid.
11 Some examples of competitions with this goal in mind include the Wellcome Trust Arts Award (https://wellcome.ac.uk/), the VISA Artificial Life International Award (https://vida.fundaciontelefonica.com/en/) and the FASEB bioart competition (http://faseb.org/).
12 Bio Art & Design Award website: http://www.badaward.nl/open-call-2017/
13 Krzysztof Ziarek, *The Force of Art* (Stanford, California: Stanford University Press, 2004), 96. Ziarek writes on p. x: "Art's importance today cannot be explained simply in aesthetic or cultural terms but has to take into consideration how artworks question the technological character of modern power."

14 Joanna Zylinska, "Taking Responsibility for Life: Bioethics and Bioart," in *Ethics and the Arts*, Paul Macneill, ed. (Springer Netherlands, 2014), 194.

15 Zylinska, 195.

16 Chantal Mouffe, "Artistic Activism and Agonistic Spaces," *Art and Research* 1.2 (2007). http://www.artandresearch.org.uk/v1n2/mouffe.html

17 Ibid.

18 Iris Karuna, "*Indecisive Green* and *Windsor Yeast*: Measuring the Immeasurable in works by Barbara Belfour and Jennifer Willet," exhibition essay (Hamilton, Canada: Centre [3], 2016). http://centre3.com/exhibition/indecisive-green-windsor-yeast/.

19 Scent evoking youth memories reminds us, of course, of Proust's madeleine.

20 I owe this observation to Ted Hiebert.

.

3 Deep Woods PCR
Responding to the Challenge of BioARTCAMP

Paul Vanouse

It is not down in any map; true places never are.

— Herman Melville, *Moby Dick*.

The Challenge

BioARTCAMP was an anachronistic, relational, and durational micro-theater in the form of an artist residency. Jennifer Willet invited six artists and two scientists to embark on a fully funded expedition to an encampment where we would set up a field station to undertake bioart projects. It was anachronistic and theatrical in that the year was 2011, not 1792 when Sir Alexander Mackenzie undertook his expedition across the continent unknown to European geographers. However, taking place in the heart of the Canadian Rockies, Willet's proposition challenged us to perform an exploratory gesture, not simply to re-explore this grand and well-preserved wilderness, but to explore an artistic project in the context of this powerful and iconic landscape. Or at least this is how I understood the challenge when I signed on.

During the residency, I produced an artwork—an extended performance—centered on restaging one of the most important biotechnological discoveries of the late twentieth century. However, the idea of exploration required unpacking before this or any art project could be formulated.

In canonical Romantic exploration narratives, the journey into the unknown is often a vehicle for self-discovery as the external environs force an internal revelation. For instance, Ishmael, the protagonist in Herman Melville's *Moby Dick*, asks that we "consider them both, the sea and the land; and do you not find a strange analogy to something in yourself?"[1] For Ishmael, the journey is a catalyst for self-reflection and the landscape is like a mirror of self. Yet Marlow in Joseph Conrad's *Heart of Darkness* dismantles poetic reflection and exposes the colonial exploit when he asserts that "the conquest of the earth, which mostly means the taking it away from those who have a different complexion or slightly flatter noses than ourselves, is not a pretty thing when you look into it too much."[2] For native peoples, the Banff region was a sacred place where medicines were gathered and healing sought in the hot mineral springs. Soon after the Banff National Park was established in the late 1800s, aboriginal peoples were excluded and traditional hunting and gathering were prohibited.[3]

Biotechnological History: The Polymerase Chain Reaction

I imagined an artistic project that would relate the romantic myth of scientific discovery narratives with those

Paul Vanouse, *Performing Deep Woods PCR at firepit with racoon*, 2011. Performance photograph, Banff National Park, photo credit: Jeanette Groenendaal.

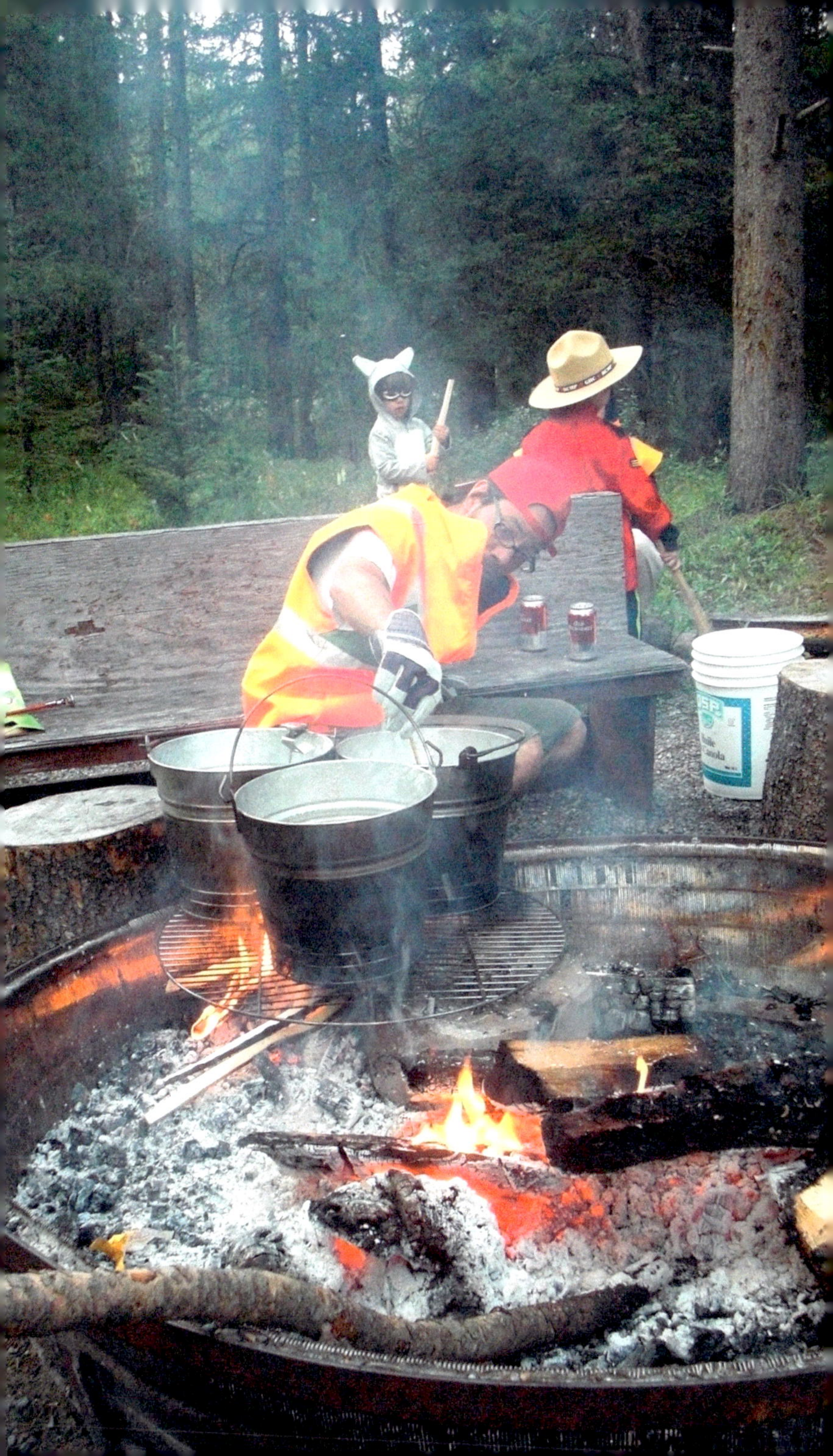

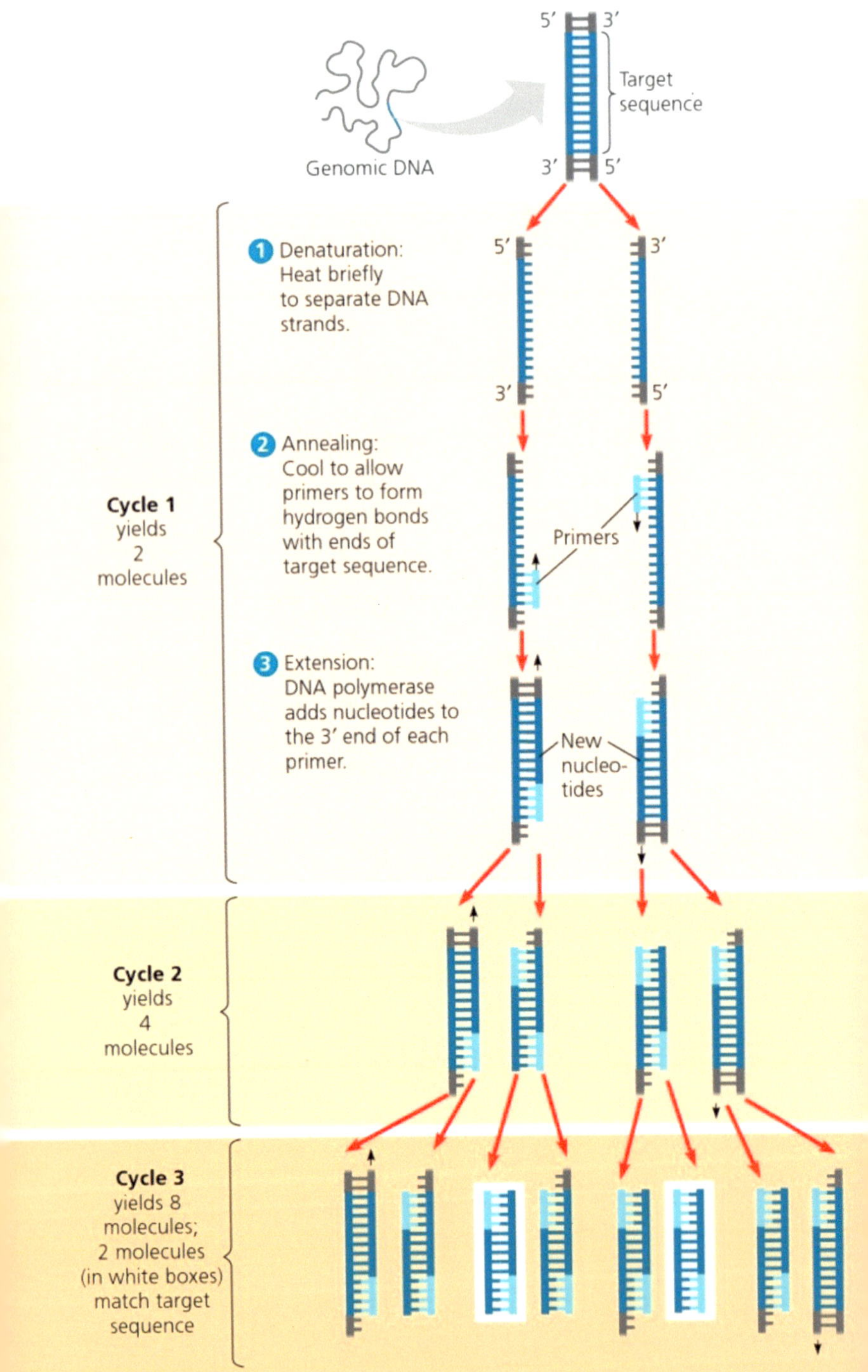

5′
3′
Target sequence
Genomic DNA
3′
5′
Cycle 1 yields 2 molecules
1 Denaturation: Heat briefly to separate DNA strands.
2 Annealing: Cool to allow primers to form hydrogen bonds with ends of target sequence.
Primers
3 Extension: DNA polymerase adds nucleotides to the 3′ end of each primer.
New nucleo-tides
Cycle 2 yields 4 molecules
Cycle 3 yields 8 molecules; 2 molecules (in white boxes) match target sequence

exploration narratives; something with similar affordances for self-discovery; something that also allowed me to reflect or subvert the colonizing tendencies of the sciences. I chose to focus on one process in particular, the Polymerase Chain Reaction (PCR)

The invention of the Polymerase Chain Reaction in 1980 has romantic, mythic dimensions and has been portrayed by its inventor, Kary Mullis, as a narrative of scientific discovery intertwined with self-discovery. The epitome of a scientific pioneer, possessing an irreverent and mischievous curiosity, he grew beyond his area of expertise to invent a revolutionary biotechnology. After completing his Ph.D. in biochemistry in 1973, Mullis began work at the Cetus Corporation to chemically synthesize oglionucleotides (short sequences of DNA), however he became fixated on an aspect of DNA replication that others had never considered. While the oglionucleotides had been used as radioactive labels to stick to corresponding regions of DNA, or as starting points to induce DNA replication, Mullis imagined using two oglionucleotides like bookends, and to allow the cells' own enzymes to exponentially copy the region. Mullis credits self-experimentation with LSD during his graduate work in the Bay area as a fundamental mind opening exercise, alongside his interest in computer algorithms and fractals, without which he may not have conceived of what he coined the Polymerase Chain Reaction.[4]

The Polymerase Chain Reaction is an elegant algorithmic process that allowed Mullis to copy a small

Dhurba Giri, *Diagram illustrating the PCR Process*, 2015.[5]

region of DNA billions of times, thereby "amplifying" this small region simply by cycling through the temperatures of a chemical reaction. The process works by first filling a tiny tube with a few microliters of: (A) purified "template" DNA; (B) oglionucleotides of about twenty bases in length; (C) chemically synthesized individual bases, A,C,G and T, called "DNTPs"; and (D) a temperature activated DNA polymerase. When heated to 95°C (Step 1) the DNA is denatured, meaning that the double stranded DNA separates into two strands, at which point copying can begin. When the temperature is reduced to about 65°C (Step 2) the oglionucleotides stick to their complementary sequence on the purified DNA. Then, when heated to 72°C (Step 3) the polymerase copies the region between oglionucleotides by assembling the DNTPs to match the template string. Then, when heated again to 95°C, to begin the cycle a second time, the strands denature and the primers release from the copied strand so the process can be repeated. DNA is replicated exponentially. Within a few years of the invention, the entire process could be performed by a PCR machine, or thermocycler, which effectively black-boxed the technology.

PCR has been utilized as a technology of "self-discovery," but also, less optimistically, as one of authoritarian identification. PCR has replaced earlier DNA typing methods in many legal cases as well as in massive government DNA databasing schemes, such as the American FBI's Combined DNA Index System (CODIS) project. It is the link to identification that particularly interested me in the context of Romantic self-discovery. However, PCR has also been used to clone DNA for DNA sequencing, gene cloning and genetic engineering as well as a myriad of related molecular biology applications.

Kary Mullis, innovator and Nobel Laureate has been described as an impulsive and opinionated oddball who alienated others at Cetus. Varied accounts have Mullis threatening other workers: sometimes carrying a gun and even punching a co-worker at an office party. In the end, he is credited with a big idea that was before its time, one

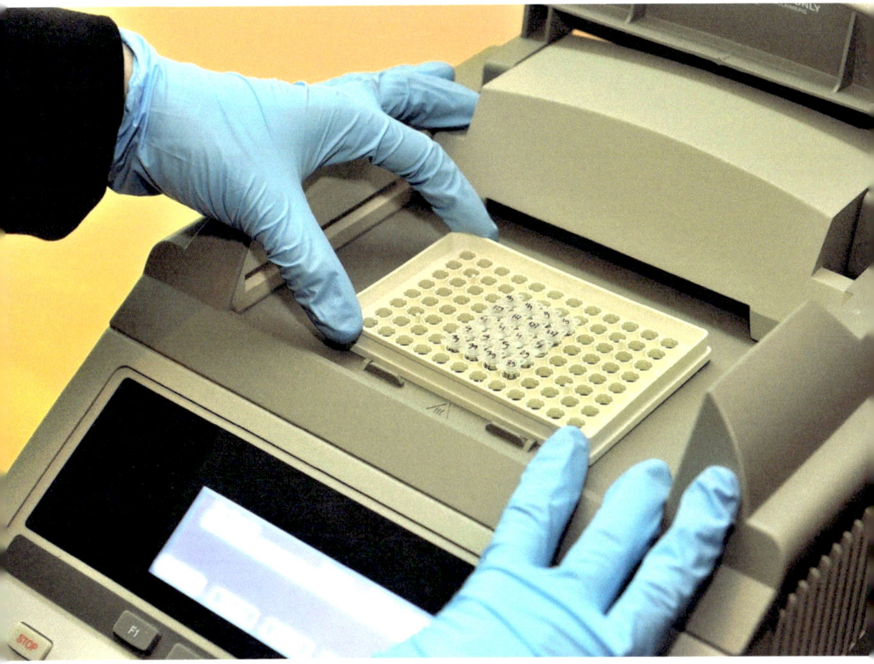

Paul Vanouse, *Typical DNA Thermocycler from approximately 2010*, 2011. Color photograph, photo credit: Axel Heise. Ernst Schering Foundation, Berlin.

that he doggedly pursued despite his other duties at the corporation. For his discovery, Mullis, along with Michael Smith, was awarded the Nobel Prize in Chemistry in 1993. However, the PCR process—the first great patent of the biotechnological revolution—was prematurely sold by

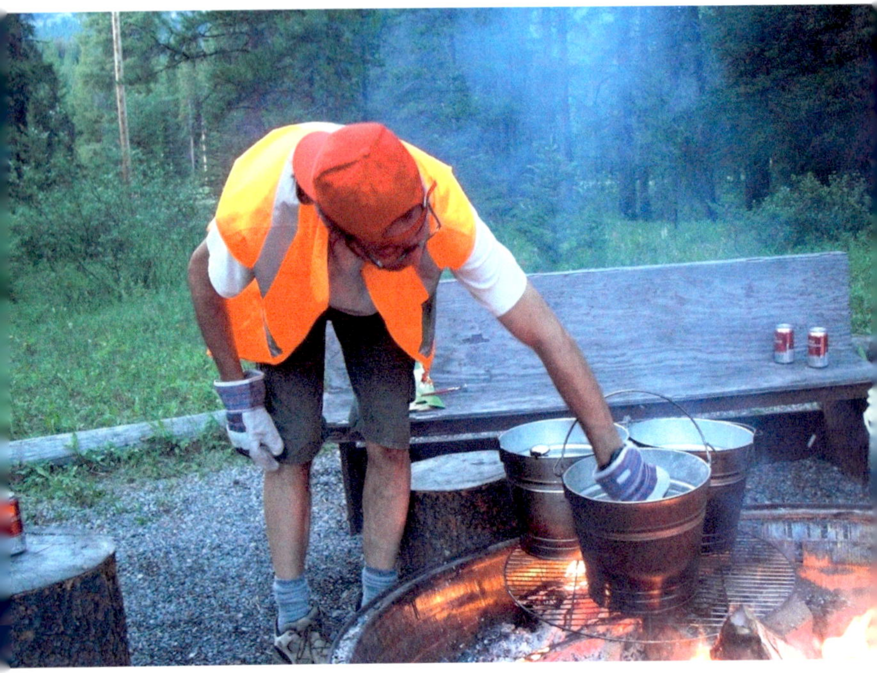

Paul Vanouse, *Performing Deep Woods PCR*, 2011. Performance photograph, Banff National Park, Canada, photo credit: Jeanette Groenendaal.

Cetus to Hoffmann-La Roche, the corporation which made PCR one of the most profitable biotechnologies of the twentieth century. [6]

Deep Woods PCR

I carried out all phases of the "high-tech" PCR process in the unlikely setting of the deep-woods using stone-age technologies such as the fire pit and three buckets of water instead of a PCR thermocycling machine. My intention was to explore the meanings of the PCR discovery/patent by using a low-tech PCR experiment, a technologically anachronistic reenactment, in the middle of the national forest. The performance conflated the scientific pioneer with a romance of the expedition and probed the idea of "self-discovery" in its widest sense. The full process follows:

Step 1: With the help of Jennifer Willet, who negotiated with the parks department for permission, I collected water from the famed sulfuric hotsprings of the region in order to incubate and extract *Thermus aquaticus* bacteria. This thermophilic organism produces an enzyme, *Taq* polymerase, which can copy DNA at the high temperatures essential to PCR's commercial effectiveness. It was discovered in geysers at Yellowstone National Forest in 1969 and intensely commercialized. In 1989, *Science* magazine named *Taq* as its first "Molecule of the Year." Presently, most commercial *Taq* polymerase is not actually produced inside *Thermus aquaticus* bacteria, but is produced by inserting a recombinant plasmid containing the cloned *Taq* polymerase gene into E.coli cells, incubating the cells and extracting the polymerase.

By attempting to harvest and utilize the original species rather than the patented gene, I sought to assert the microbe's nature rather than its colonized product, while also recalling the history of aboriginal medicines gathered from the hotsprings prior to intensive colonization.

Step 2: I extracted my own DNA using an educational kit containing *Instagene Matrix*, from the Bio-Rad company. The extraction protocol involved swishing saline in my mouth for 60 seconds and dispensing into 2 ml capped tubes, adding *Chelex* resin beads (which bind to non-DNA components of cells in the spit sample), heating to actually lyse the cells, then collecting the supernatant liquid containing my DNA.[7] Then I combined—in a 2 ml tube—minute quantities of DNA, Taq, DNA primers designed to target the *Alu 92* gene, DNTPs, and water.

Step 3: I performed the experiment in pioneer fashion, without a PCR machine, but rather by using a campfire. It required that I keep water buckets at precise temperatures: 95°, 65°, and 72° Celsius. Then I "thermocycled" the buckets—switching between them 120 times, or 40 cycles of about 90 seconds each. I was simultaneously maintaining water temperatures by adding wood to the fire and shifting bucket placement. This stage was the focus of *Deep Woods PCR* and what I considered the artistic and scientific challenge. It took hours.

Filmmakers Zoot Derks and Jeanette Groenendaal tirelessly documented the event in its entirety. Angus Leech premiered an original composition, *Thermocyclin' 'neath the moon*, which he sang as he strummed his acoustic guitar. Other artists and scientists from the *BioARTCAMP* also joined me at the campfire for the all-night performance, including: Jennifer Willet, Tagny Duff, Marta de Menezes, Adam Zaretsky, Marie Pier Boucher, Kurt Illerbrun, Iain and Louise Chance-Baxter&, Bulent Mutus, Grant Yocom, Brit Wray, Jamie Ferguson, Dave Dowhaniuk and Kacie Auffret.

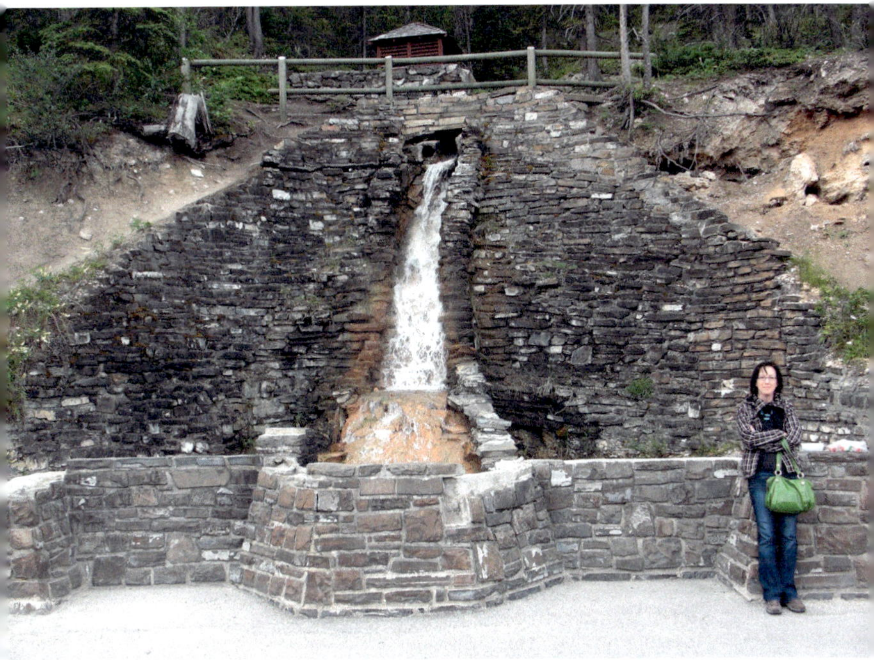

Paul Vanouse, *Jennifer Willet at hotspring in Banff National Park, Canada*, 2011. Color photograph.

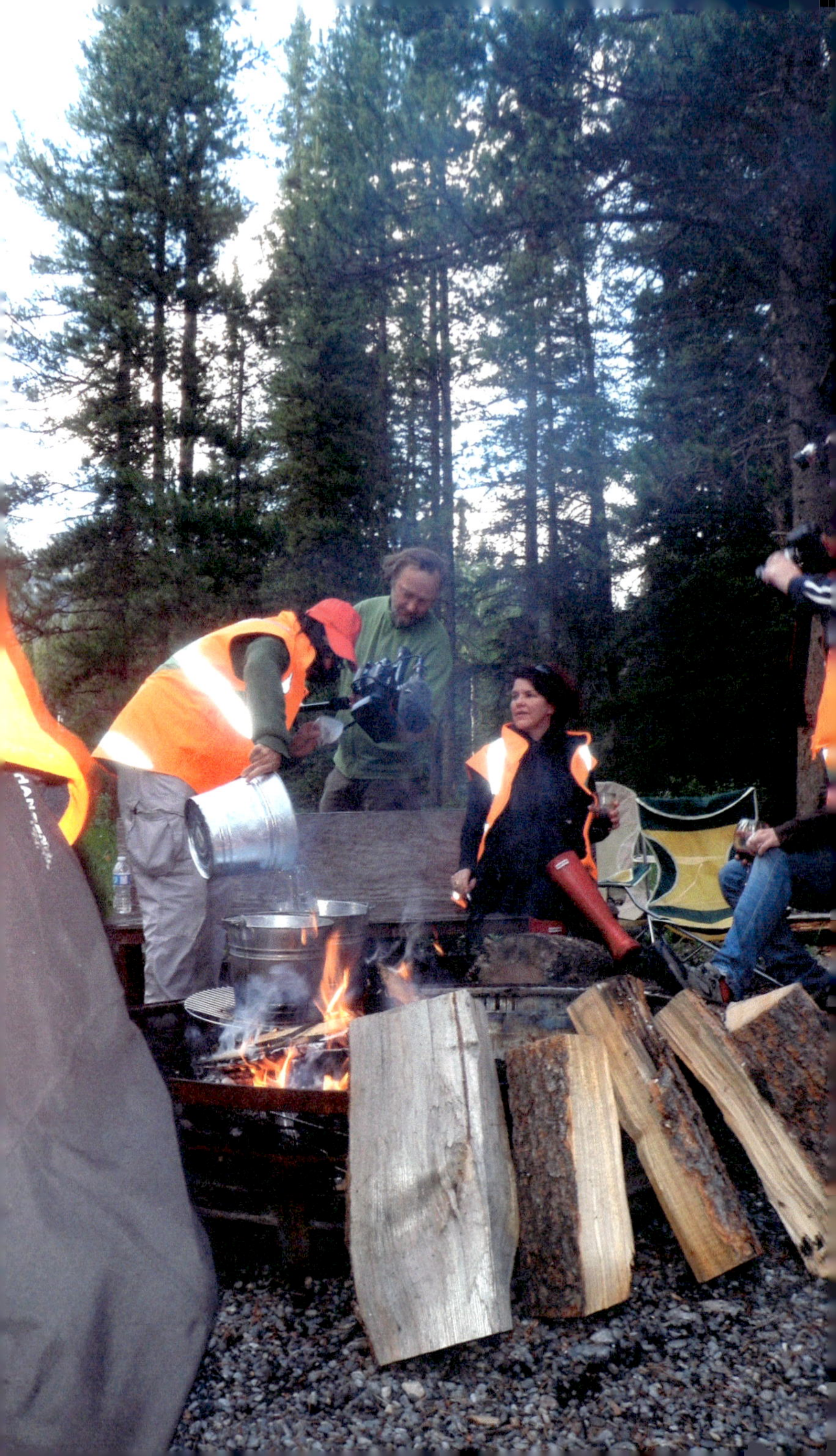

During thermocycling, I attempted to channel Mullis. Rafael Vanouse played the part of a fluorescent green raccoon and repeatedly chanted "amplify the source, not the signal," a phrase that refers to the genius of PCR for DNA analysis. Prior to PCR, scientists seeking to visually analyze DNA fragments needed to radioactively label target sequences and typically they sought better imaging apparatus or stronger radiation to enhance their dim images. Mullis, on the other hand, realized that DNA regions could be amplified exponentially by successively initiating the DNA transcription process and thus greatly enhance any imaging or analytical process. In his autobiography, Mullis describes a benevolent encounter with a fluorescent green raccoon muse that greets him with "good evening, doctor" near his cabin in Northern California. In my rendition, the raccoon is a benevolent trickster who helps Mullis on his journey of discovery.

Step 4: To judge if the PCR worked, I ran the samples in a DNA electrophoresis gel. Miraculously, it succeeded. DNA bands are clearly visible in the second and third lanes of the gel (the fuzzy bands are probably a primer smear, not DNA, but the crisper bands just below are DNA), which means *my DNA was amplified.* Furthermore, these DNA band locations show I am heterozygous for the PV92 *Alu* gene. This genetic site is often used for genotyping, as there is variation across our species. I was not expecting the process to actually work, since PCR reactions are very sensitive and tricky even in laboratory conditions.

Paul Vanouse, *BioARTCAMP participants at the fire pit*, 2011. Color photograph.

Conclusions

Just as context is meaningful in any cultural activity (such as an artistic performance) the context in which PCR was "invented" by Mullis is fundamentally linked to its cultural meaning, economic value, and scientific merit. When Mullis began his exploration, there was a big problem. It had no purpose. The idea of duplicating a region of DNA exponentially was "cool," but of little use in research or industry. However, by the time he obtained publishable results, the technology had been shown to have great utility for genetic engineering (allowing one to duplicate a gene for transplant) and DNA typing (allowing one to compare highly variable regions of DNA). Changing context turns a neat trick into an indispensable laboratory tool.

Mullis's "invention" occured at the brink of the biotech era, while patent laws were being stretched and as private biotech corporations outmaneuvered the university research labs. "Invention" is used in quotes here because every part of this process had already been invented—PCR patents were highly contested by the scientists who had developed the individual components of the PCR process. PCR is only an invention in the changing context of research science, which made Mullis's "idea" unique.

In this sense, perhaps the deep woods context of my own performance does more than highlight the complex play of humans, non-humans, and the context of technological invention—often called the "actor-network model." Hopefully this amateur using primarily stone-age technology to perform contemporary molecular biology playfully opens doors to other non-specialists seeking informal, non-outcome-driven ways to participate with the techno-sciences.

Postscript

While conceiving *Deep Woods PCR* before *BioARTCAMP*, I had a simplistic conception of Mullis. He was the famed biochemist who could be performed and parodied because he was a public figure with unconventional libertarian opinions and conducts incongruous with societal conceptions of a serious scientist. When I returned from the experience, I reinvestigated and reimagined Mullis's

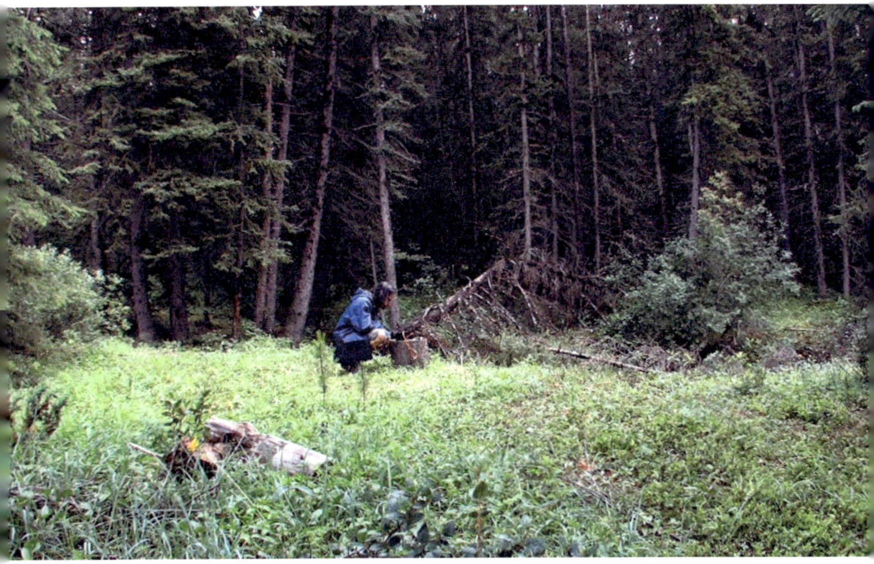

Paul Vanouse, *Performing Deep Woods PCR*, 2011. Performance photograph, Banff National Park, Canada, photo credit: Jeanette Groenendaal.

role in the 1995 O. J. Simpson trial, in which he was in the courtroom as an expert witness for Simpson, but never called to the stand. According to Mullis, the prosecution team had been tailing him outside the courtroom and planned to use

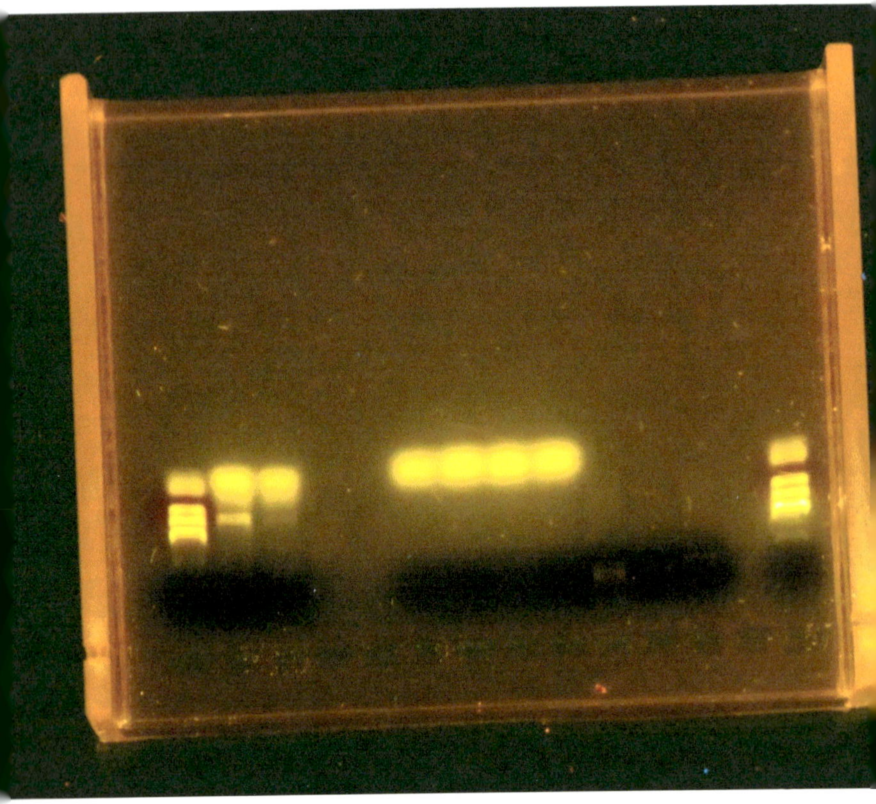

Paul Vanouse, *Final image of DNA gel from Deep Woods PCR, Banff National Park, Canada*, 2011. Color photograph of electrophoresis gel.

his advocacy of LSD to discredit his testimony. As a Nobel Laureate who invented the technology used in many DNA tests, his scientific opinions would otherwise be influential. Mullis thought the idea that the lab doing the DNA analysis was literally run by the prosecution violated standards of objectivity as well as the "first principles" of scientific method.[8]

In 2013, Dr. William Thompson, a member of the "Dream Team"—Simpson's defense team led by Johnny Cochran—relayed to me an anecdote about another idea of Mullis's: to invent a personal privacy spray called "DNA Anonymous" that could be used to mask one's DNA fingerprints. This led me to imagine what might have transpired in the O. J. Simpson trial had Mullis been called to the stand. While the DNA evidence in the trial had been thrown out by Judge Lance Allen Ito because of *chain of custody* missteps by the investigators, it would seem Mullis's testimony might have been more faultfinding of DNA evidence and assumptions as a whole. For instance, perhaps a new precedent for the metrics of certainty would have been reached, or adversarial participation in the laboratory work would have been mandated. Simpson's defense team accomplished the task of a legal defense, to advocate for the accused, whereas perhaps Mullis on the witness stand would have deconstructed DNA forensics for the broader good of the justice system.

Later, in 2014, I visited with Mullis and his wife Nancy at their modest Southern California home. While some biochemists in the 1980s objected to the Nobel accolade for PCR being given to Mullis, he reaped little of the big financial rewards. While the Roche corporation made billions of dollars from PCR, Mullis received only a $10,000 bonus for his achievement from the Cetus corporation at which he worked. While this, in addition to his half share of a Nobel Prize and speaking fees throughout the nineties, would seem to provide for financial security, the Mullises were not barons of biotech billions. Kary Mullis now seemed more like a (retired) actor in a powerful biotechnological apparatus with replaceable parts. This awareness has underscored my purpose and understanding of *Deep Woods PCR* as a performative exploration of actors and actants, humans and non-humans, protocols and patents, rather than simply as a critique or a parody.

Post postscript

Since I have no doubt strained the conflation of scientific, terrestrial, and self-exploration, as well as pushed the analogy between Ishamael and Mullis's explorations and my own performative artwork, I must also prevent the reader from carrying these analogies too far. Specifically, while Ishmael has his Ahab and Mullis his Cetus corporation, my journey with Jennifer Willet was one of equity and collaboration. Willet is not Ahab in this anachronistic imaginary land in which *Deep Woods PCR* was enacted. *BioARTCAMP* however, may well be analogous to a mind opening psychoactive drug, which might be a more suitable comparison.

Notes

1 Herman Melville, *Moby Dick* (New York: Random House, 1967), 293.
2 Joseph Conrad, *Heart of Darkness* (New York: Dover Publications, 1990), 6.
3 Banff & Lake Louise Tourism website: https://www.banfflakelouise.com/banff-national-park/history-heritage. The website notes that "these policies have been reversed over the last 50 years."
4 Kary Mullis, *Dancing Naked in the Mind Field* (New York: Vintage Books, 1998).
5 Image source: Dhurba Giri, "Polymerase Chain Reaction (PCR): Principle, Procedure, Components, Types and Application," Laboratoryinfo.com, July 27, 2015. http://laboratoryinfo.com/polymerase-chain-reaction-pcr/
6 Paul Rabinow, *Making PCR: A Story of Biotechnology* (Chicago: University of Chicago Press, 1996).
7 For more information on the extraction process, see the Bio-Rad webstie: http://www.bio-rad.com/en-us/product/instagene-matrix
8 Mullis, *Dancing Naked in the Mind Field.*

4 Tuesday 4:41 pm

Christian Kuras

Image Credits

Christian Kuras, *Tuesday 4:41 pm*, 2017. Ink on paper.

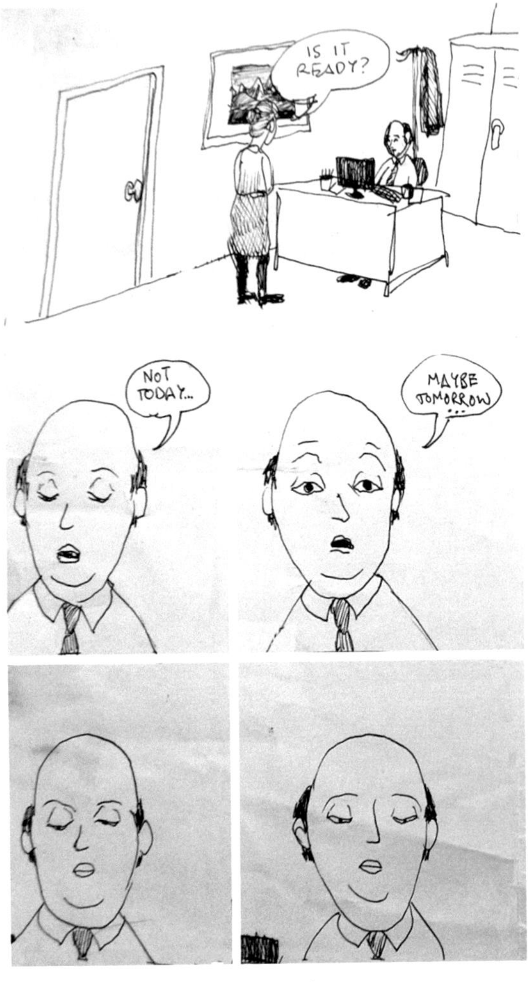
IS IT READY?
NOT TODAY...
MAYBE TOMORROW ...

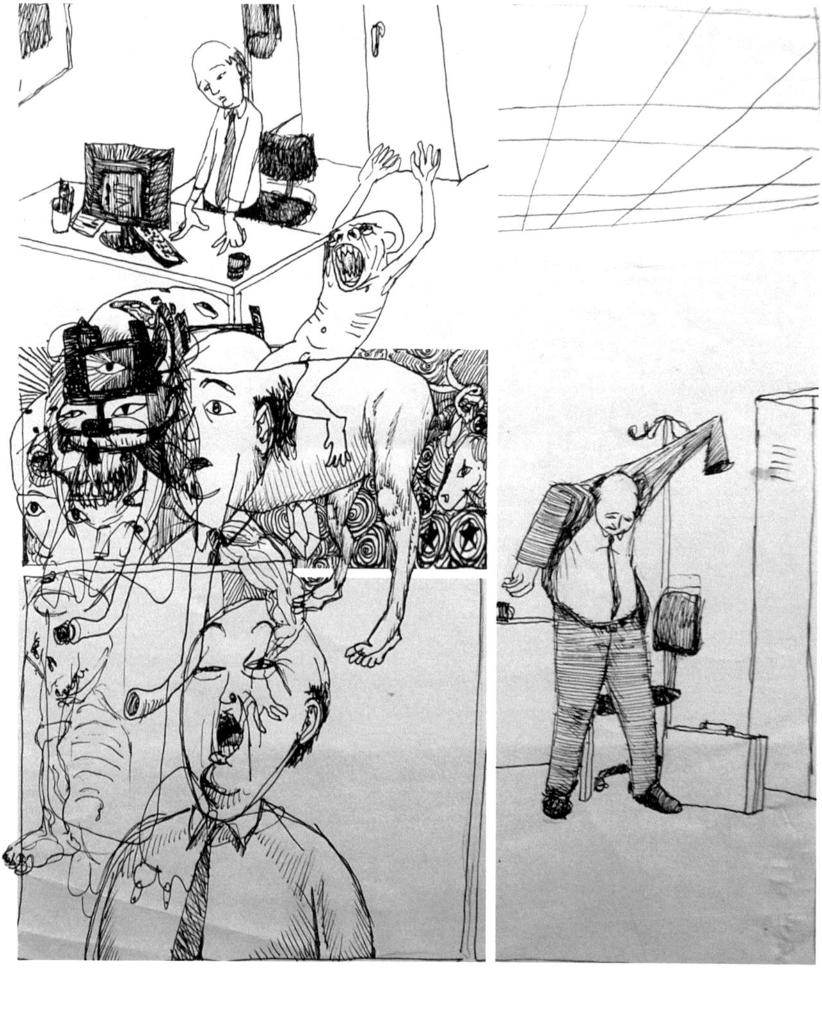

5 Untitled (BioARTCAMP)

Louise Chance-Baxter&
and IAIN BAXTER&

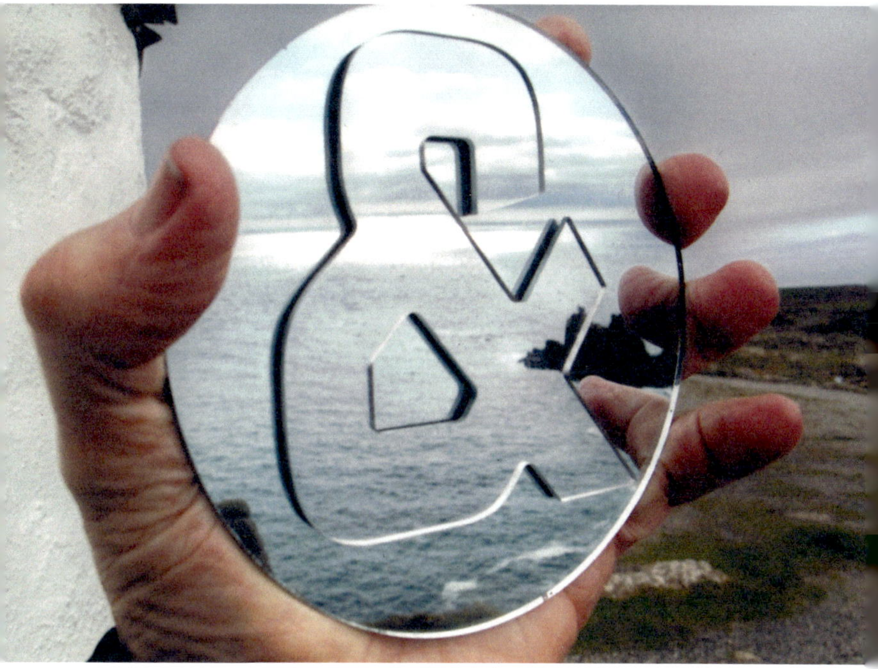

Image Credits

Louise Chance-Baxter& and IAIN BAXTER&, images from the series *Untitled (BioARTCAMP)*, 2011. Performance photographs.

Untitled (BioARTCAMP) documents a series of actions improvised by IAIN BAXTER& and Louise Chance-Baxter& during their 2011 Banff residency. This project carries forward the artists' decades-long history of collaboration, including explorations of ecology, gender, nationhood, and race, such as *Keeping Abreast of Breast Cancer* (1996) and *One Canada Video* (1992). In *Untitled (BioARTCAMP)*, the artists approached the scenic landscape of Banff—a site intimately intertwined with the biography of BAXTER& from his early training as a competitive skier—as a laboratory for spontaneous experimentation. Drawing on her early training in fashion, Louise Chance-Baxter& devised fantastical costumes with which to enact otherworldly, erotic tableaux. These imaginative interventions within the wilderness are complicated in several shots by the presence of a life-sized artist's mannequin. The ligneous materiality of this figure draws attention to the doubly commodified status of Banff, which is both a major tourist destination and a standing-reserve of unprocessed timber products. This attentiveness to the staple-commodity character of the Canadian landscape is a recurring theme in the work of BAXTER&, whose satirical gaze was sharpened by his early scientific training in ecology and zoology at the University of Idaho.

— Adam Lauder, 2017

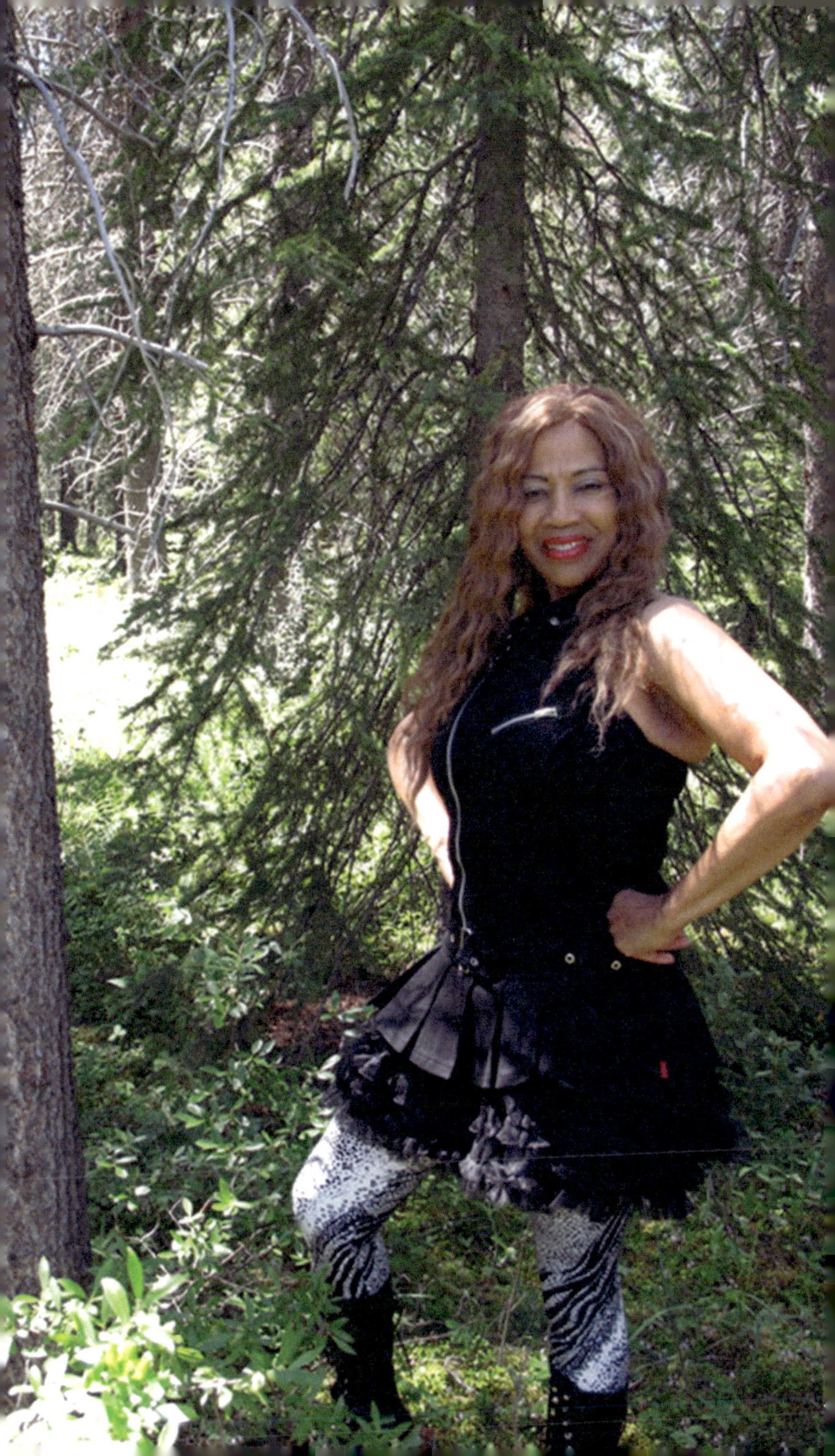

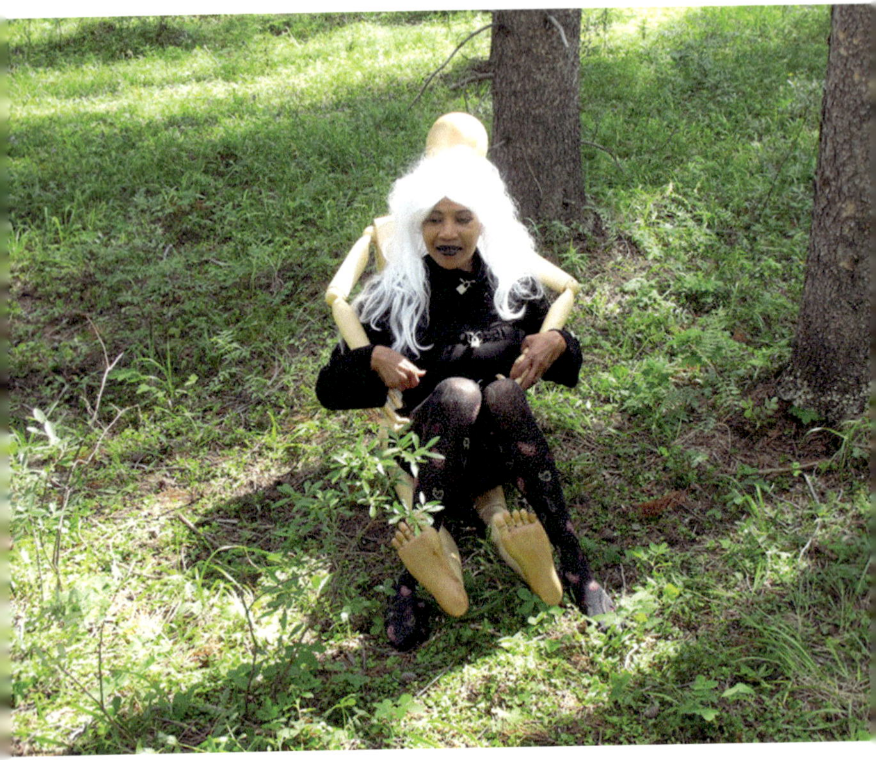

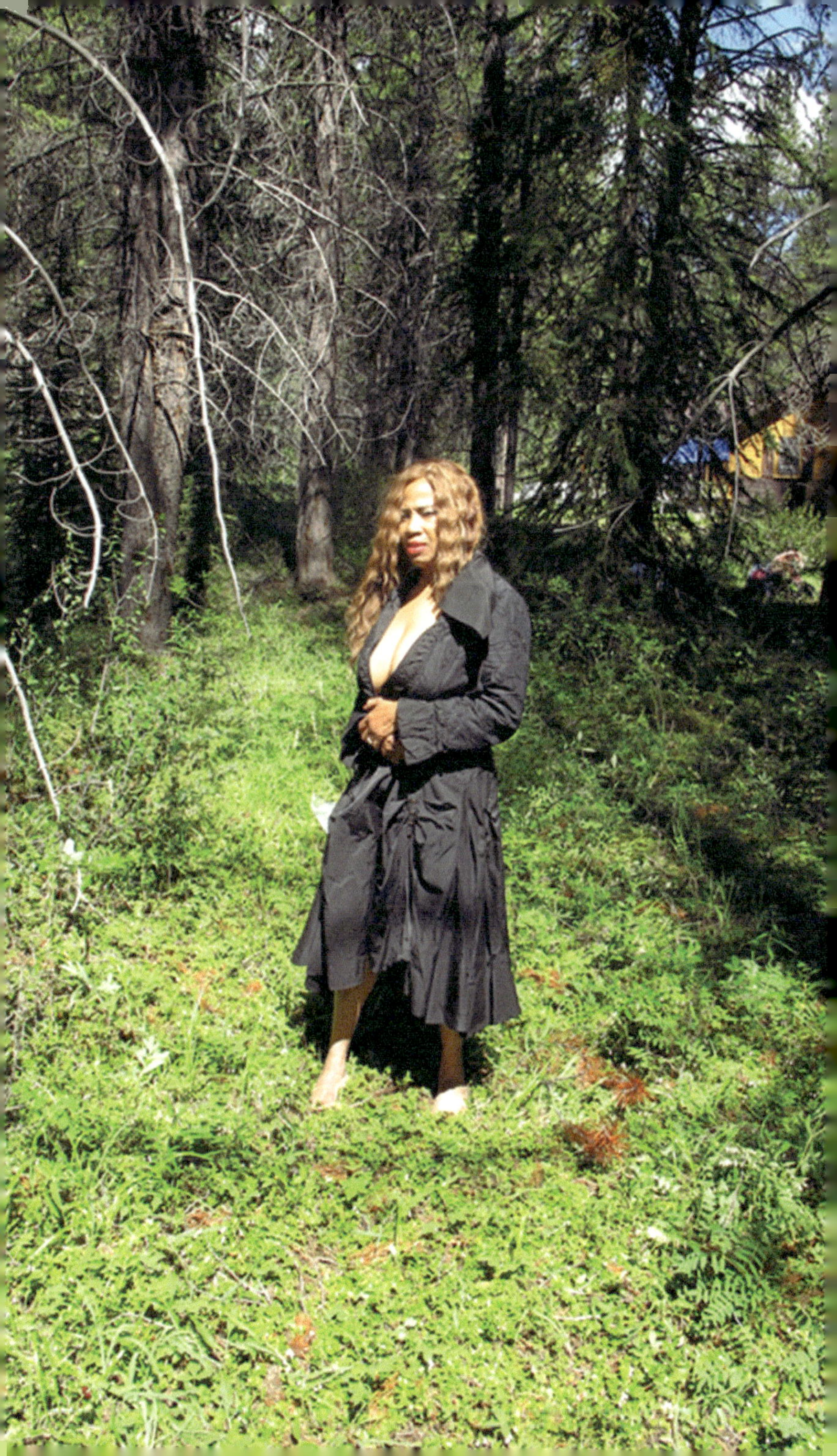

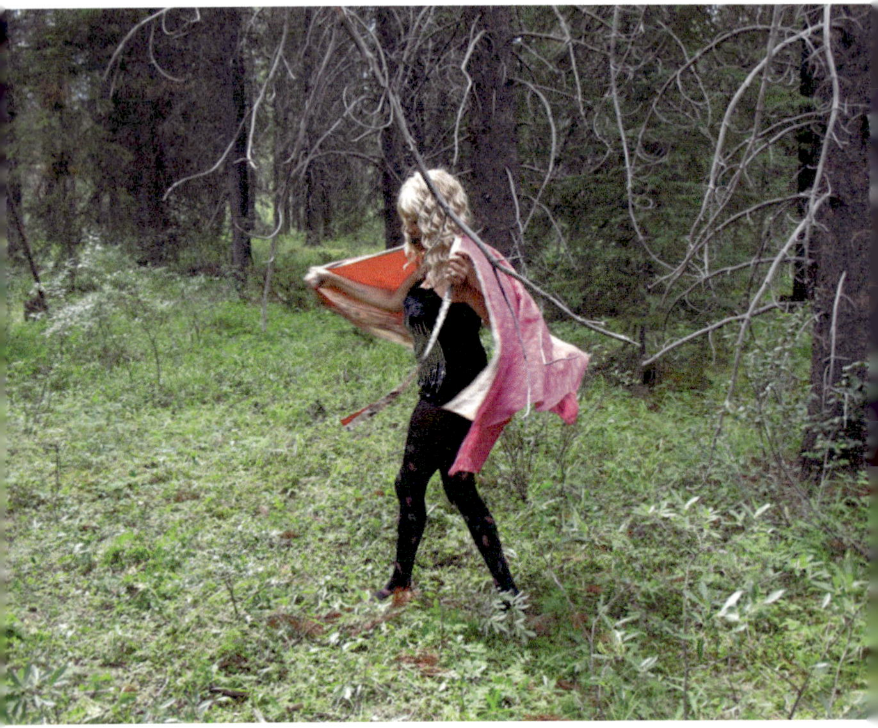

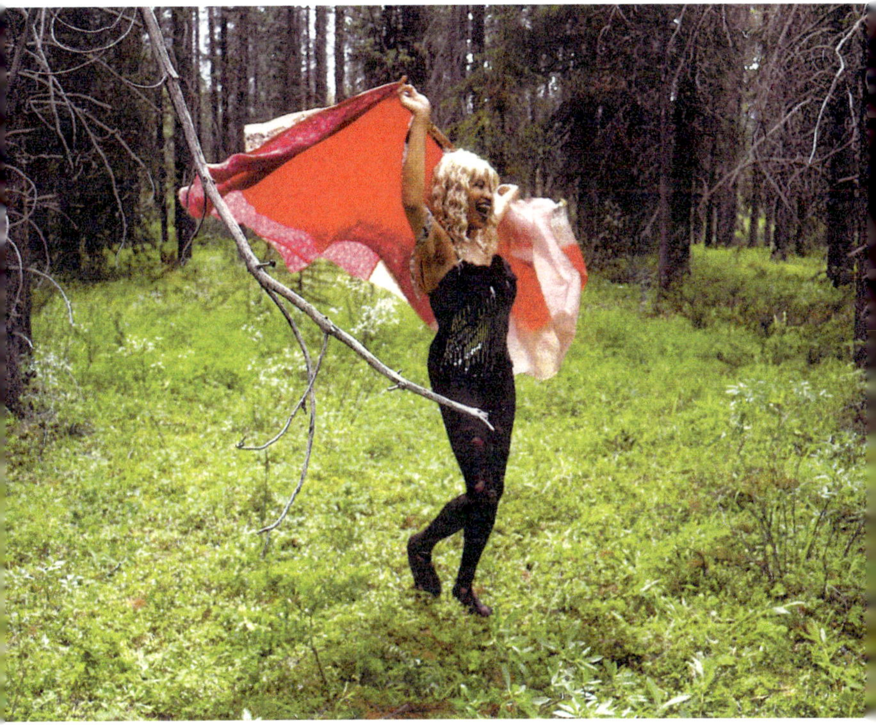

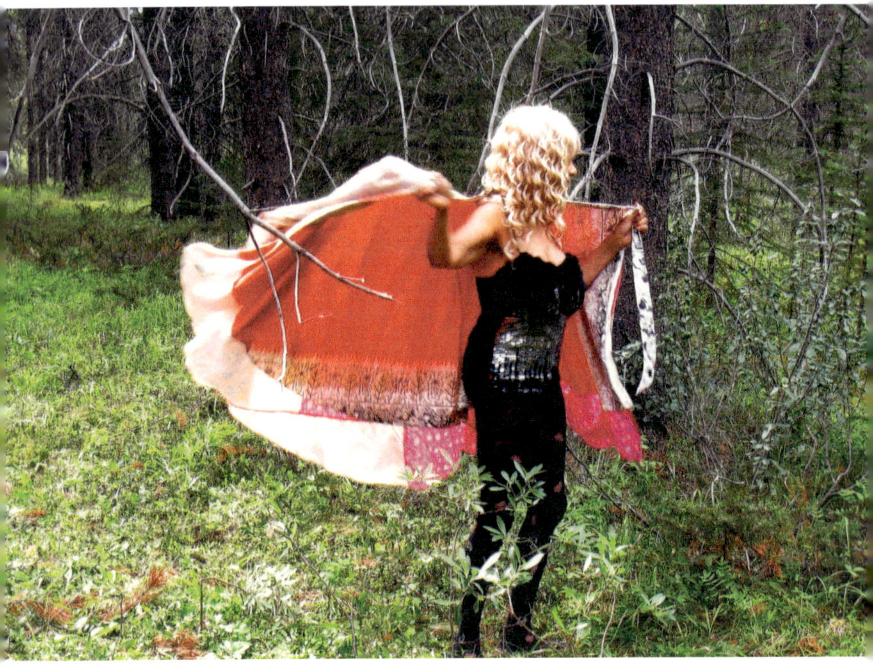

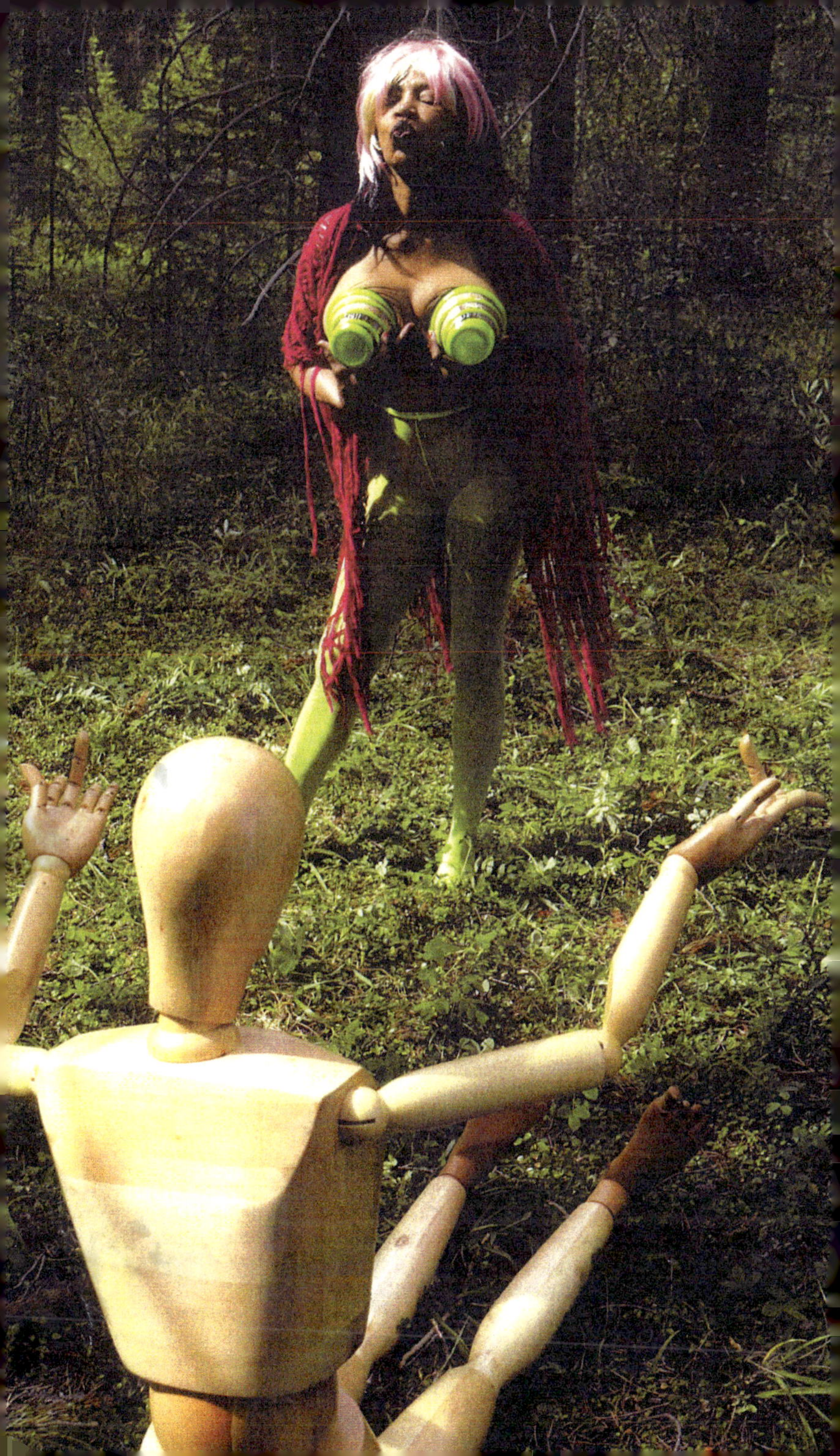

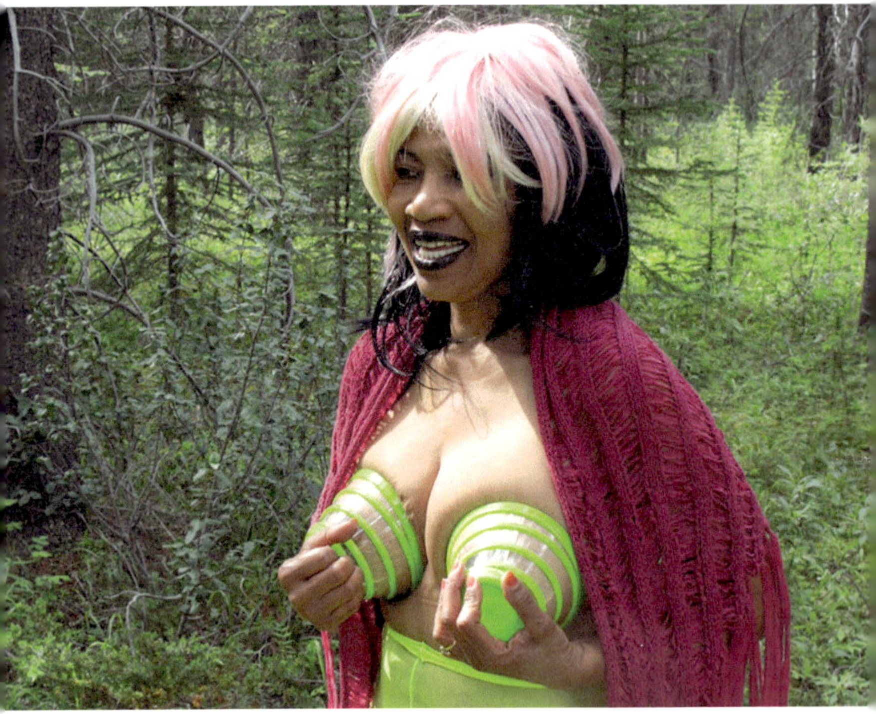

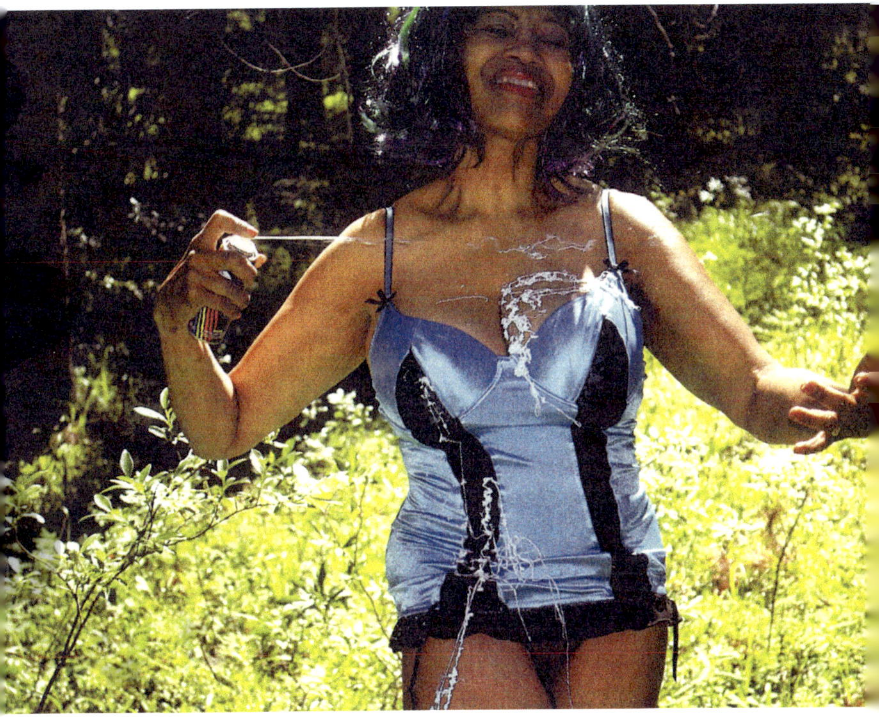

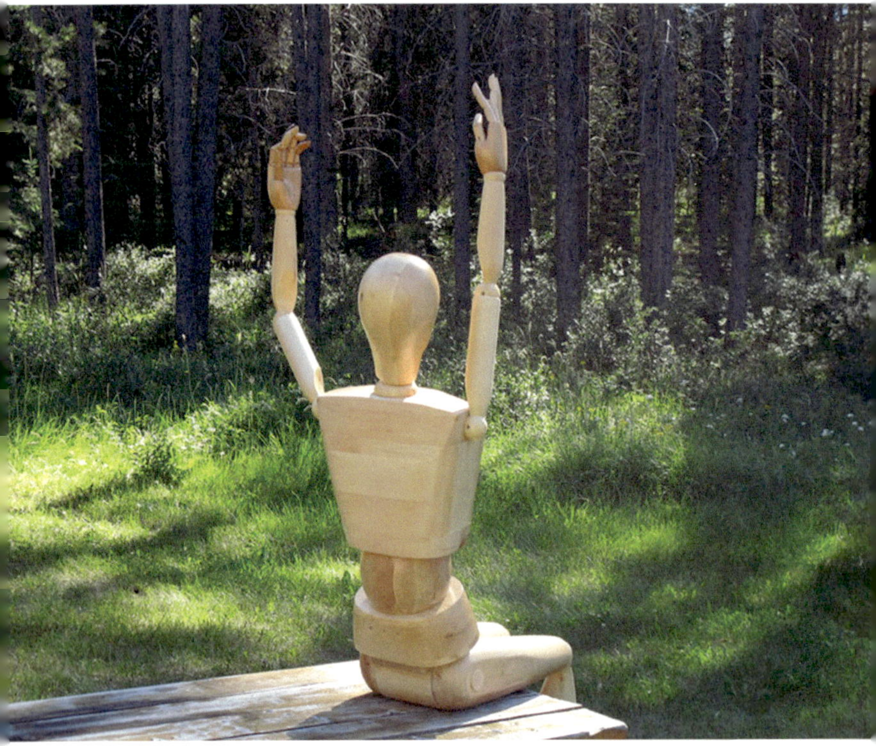

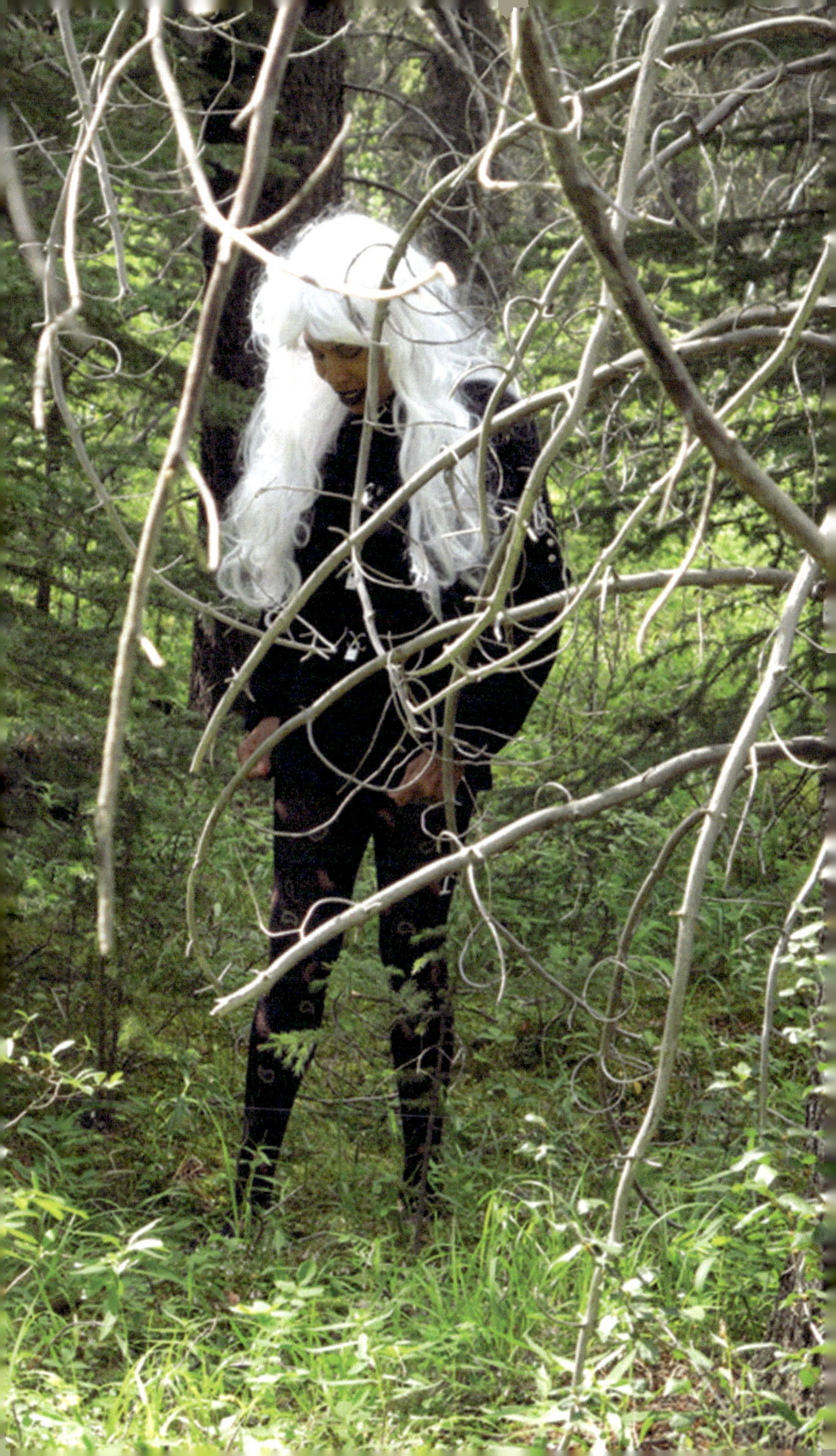

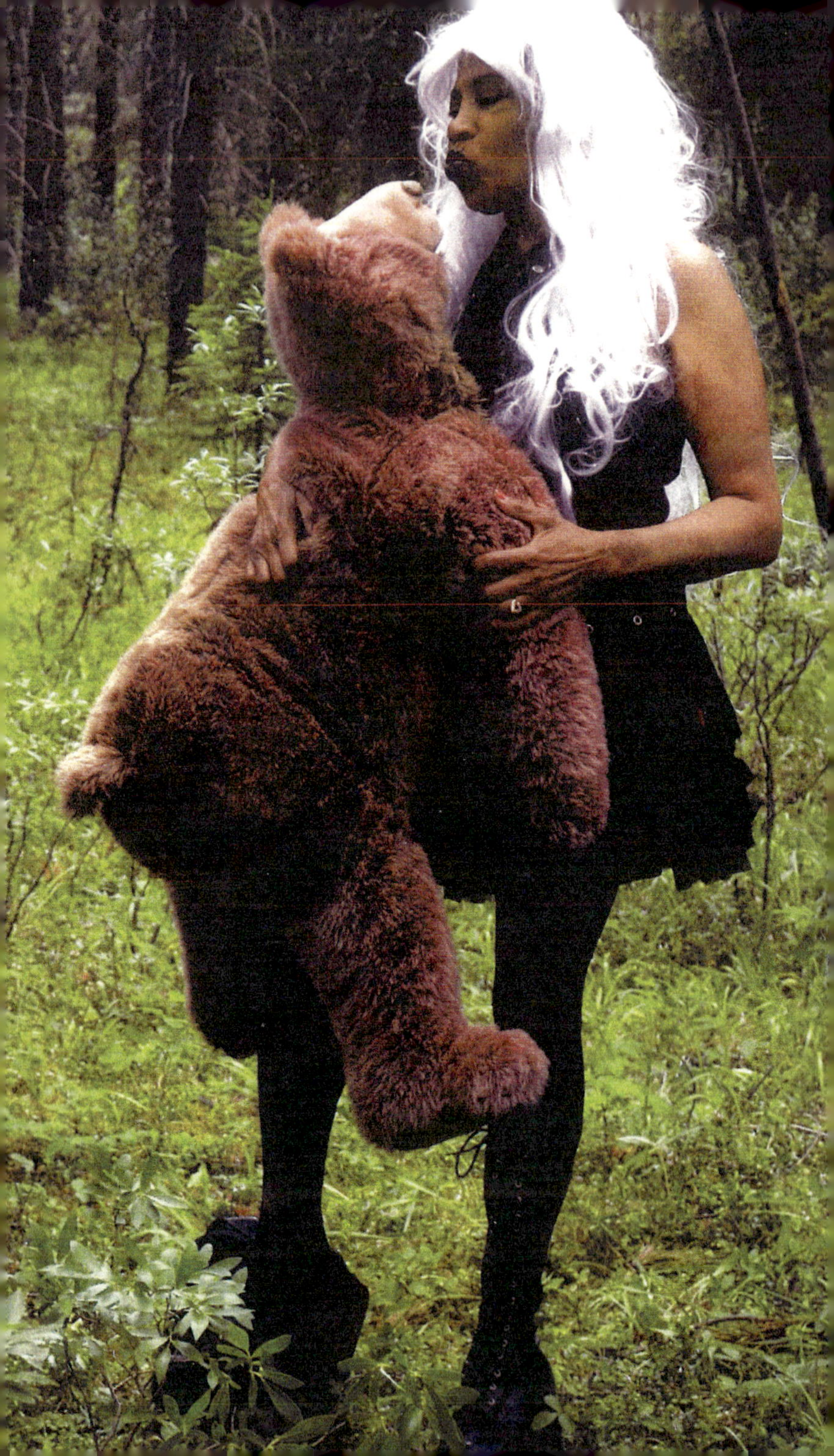

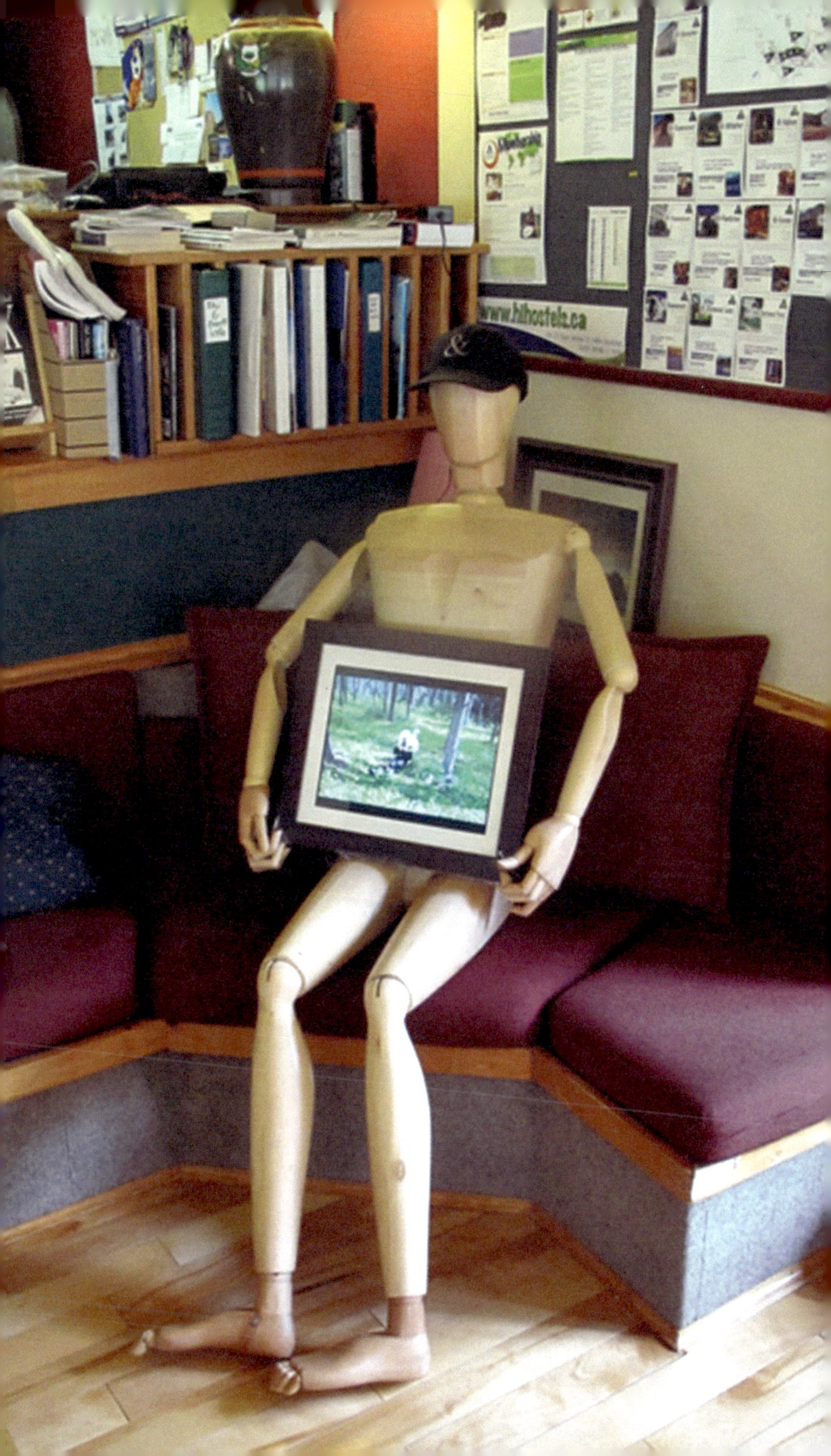
www.hihostels.ca

Acknowledgments

I wish to thank my wife Louise Chance-Baxter& for all of her love & inspiration & for working with me for over 30 years on many collaborative projects. Also, many thanks to Jennifer Willet for all her help & for inviting Louise & I to be a part of *BioARTCAMP* in Banff National Park in 2011 & Jeanette Groenendaal & Zoot Derks for their video reporting of our collaborative project. Also, Professor Earl Larrison my Zoology professor at the University of Idaho in the late 1950's who inspired me to become an artist & for inviting me to be the artist to illustrate 200 drawings of birds & animals for his 1961 publication on the wildlife of the Northern Rocky Mountains & Galen Hansen, famous American artist, for his insights and encouragement while I was a graduate student at Washington State University in the early 1960's & David P. Silcox, the André Malraux of Canada, for all his support & friendship for over 50 years & thanks to Mike Marcon, artist/designer for assisting me in designing different projects over the last few years & also Alison Zilli, artist, for editing and helping with texts & Victor Romeo, artist/illustrator, for his collaborative work over the last decade & Dan Bernyk for all his support and technical advice over the years & Julie Sando for her help with my photo archives & documentation & finally Adam Lauder for his support, insights, research & friendship. &MAN

6 Petrography
Bitumen in Place

Warren Cariou

My photographic practice begins and ends in a place made infamous largely through photographic and video imagery: the Athabasca tar sands. This region, home to some of the richest petroleum deposits in the world, has attracted enormous investment by international oil companies, which have transformed the landscape on a massive scale, completely altering or even removing entire ecosystems, as hundreds of square miles of boreal forest are scraped away in order to strip mine the lucrative oil-bearing bitumen underneath. Much of the recent attention given to this activity has been channeled through photography and video, both in documentary forms and in the more self-consciously aesthetic work of artists like Edward Burtynsky and Louis Helbig. It is the overwhelming *visibility* of tar sands mining that makes it such a compelling and horrifying subject for photography. The spectacle of such earth-work simply begs to be photographed. But questions remain about the politics of spectacular and aestheticized representations of these environmental disaster zones. Does representing these mines as essentially abstract works of land-art risk diminishing the political impact of the images? Is there a way to photograph such devastation in a more intimate and embodied way, one that resists the distancing effect of spectacle and shows us how we as viewers are connected to the damage we are beholding?

Since my home territory is near the Athabasca bitumen mining operations, I have a personal stake in these questions of representation, and I have long been struggling with the ethical and political challenges posed by the ease with which

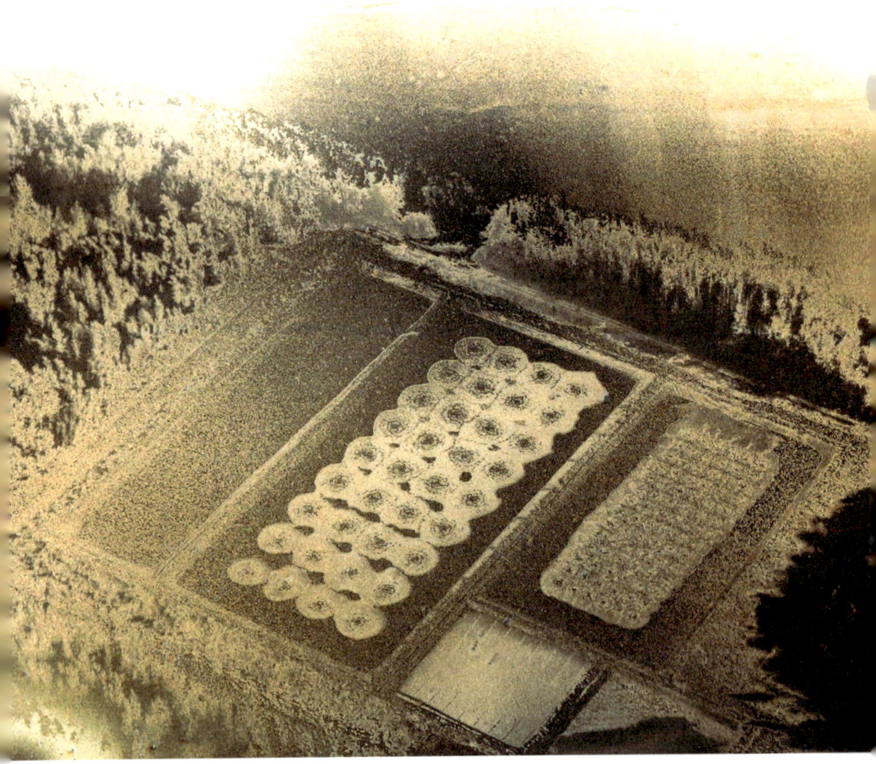

Warren Cariou, *Water Treatment Facility on Bank of the Athabasca River. V 1 of 3.*, 2014. Athabasca Bitumen and Lavender Oil on Aluminum Plate, 8.5" x 11".

these industrial landscapes become aestheticized through photography. In 2013, I shared some of these concerns with Jennifer Willet, whom I had recently met as part of a large collaborative research project on performance

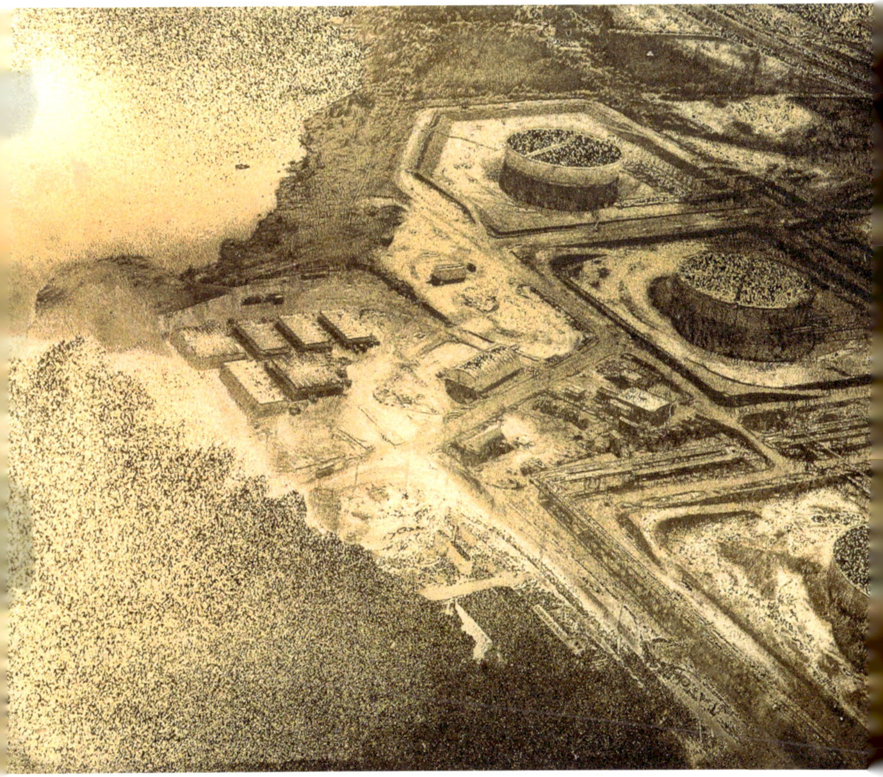

Warren Cariou, *Bitumen Storage Facility Occluded by Smoke. V 1 of 2.*, 2014. Athabasca Bitumen and Lavender Oil on Aluminum Plate, 8.5" x 11".

and politics in the Americas, and she encouraged me to experiment with the physical substance of the bitumen itself, seeking ways that it could be used as a medium of representation. After a great deal of trial and error, and after further encouragement at a bioart workshop in the summer of 2014, I was finally successful in this ongoing series of experiments. By repurposing the early techniques of Nicephore Niepce, who used bitumen of Judea to create what is now known as the first photograph, I was able to create images of these mined landscapes using the tar sands bitumen itself as the photosensitive material. These "petrographs," as I call them, are made through the action of sunlight upon petroleum, and they are created from the very land they represent.

Jennifer Willet's work, troubling the boundaries between the toxic and the natural, the contained space of the lab and the unfettered realm of the field, has been an abiding fascination for me during the course of my petrography experiments. She has taught me to be attuned to the biological and fundamentally *emplaced* nature of my newly chosen photographic medium. Like other forms of petroleum, bitumen is of course biological material, composed of the greatly-altered remnants of aquatic organisms buried in the earth for many millions of years. Petroleum can also support some forms of life; for example it is consumed by certain microorganisms, and it can be used to provide stability for disturbed soils, enabling the

Warren Cariou, *Tailings Pond Inlet with Whirlpool. V 1 of 3.* (detail), 2014. Athabasca Bitumen and Lavender Oil on Aluminum Plate, 8.5" x 11".

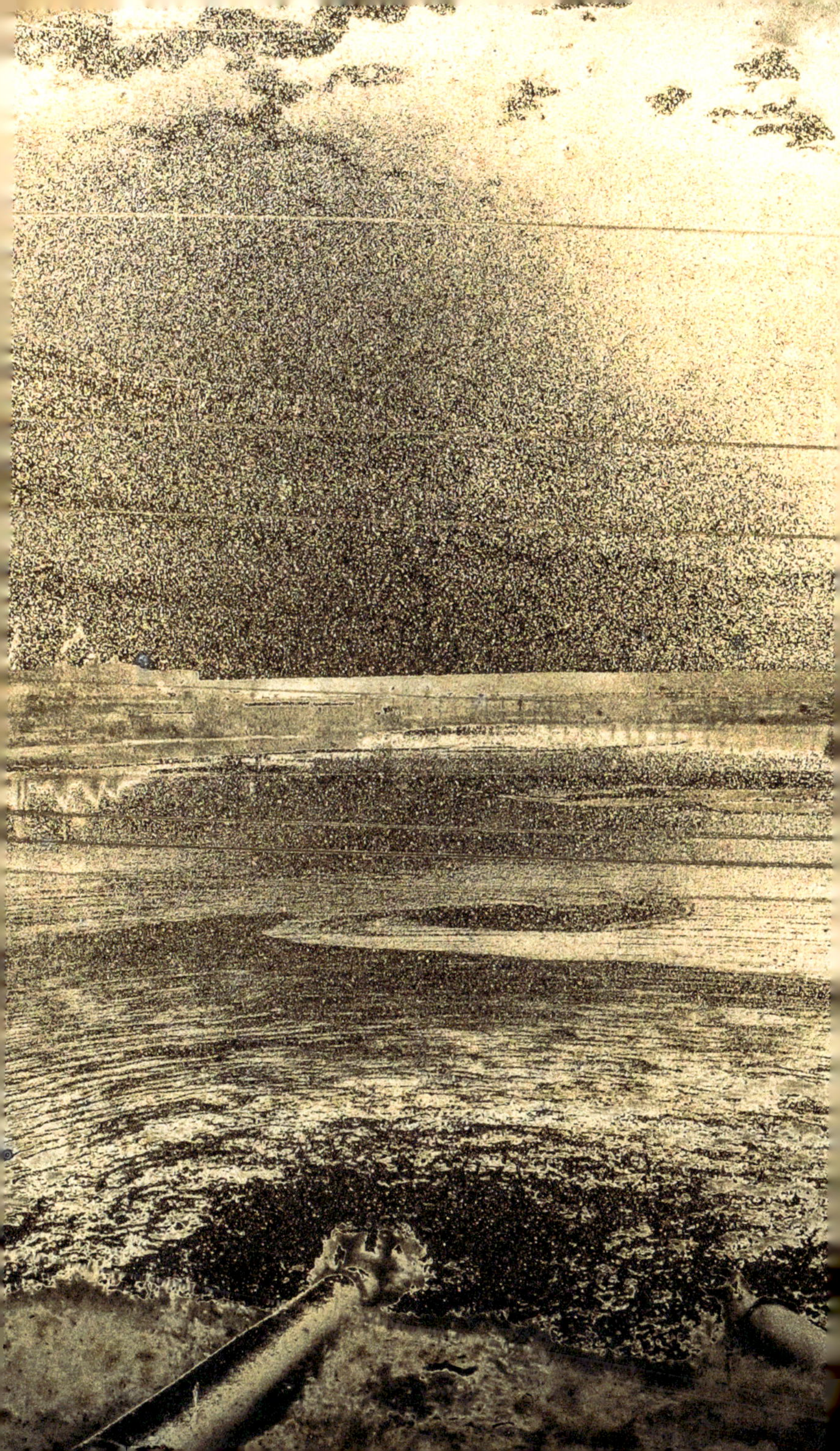

growth of plants. Nonetheless, we are also well aware that bitumen and other forms of petroleum can be toxic in many contexts. For example, the aromatic hydrocarbons that are byproducts of petroleum combustion are carcinogens, and

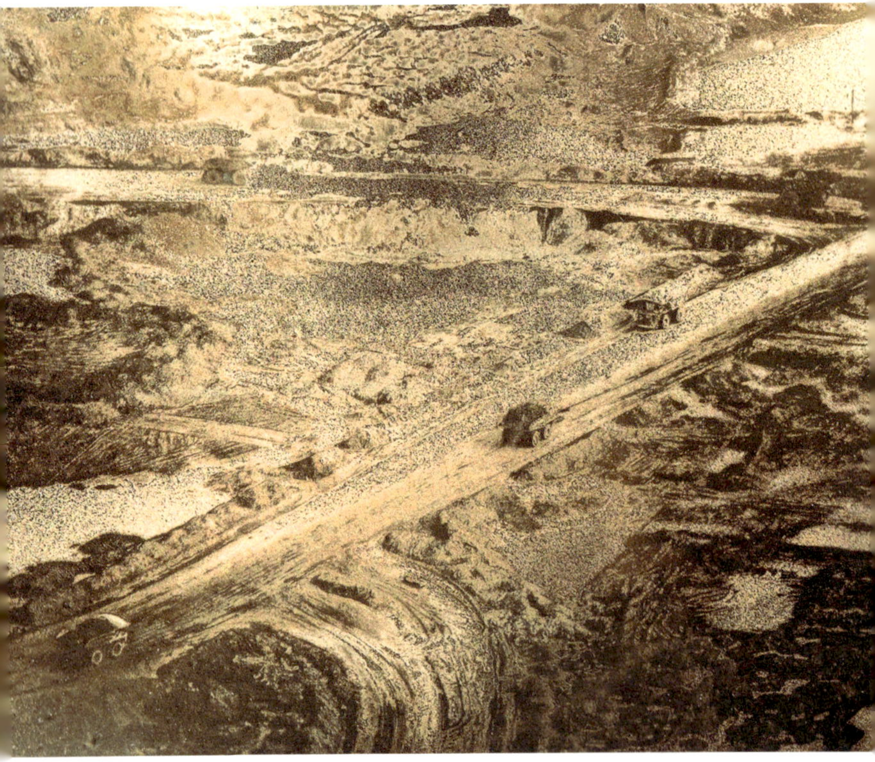

Warren Cariou, *Bitumen Mine Landscape with Five Trucks. V 1 of 2.*, 2014. Athabasca Bitumen and Lavender Oil on Aluminum Plate, 8.5" x 11".

gasoline fumes can cause permanent impairment of brain functions or even death. Oil spills cause immediate and long-term damage, especially in aquatic zones, and climate change will cause untold levels of destruction for millennia, reshaping ecosystems and rendering formerly hospitable regions dangerous or uninhabitable.

While the negative consequences of oil extraction and consumption vastly outweigh the positives, I think it is important to maintain a sense of oil's fundamental ambiguity in its relationship to life. I am fascinated by the duality of this substance, which embodies destructive and creative possibilities at the same time. My process of petrography in some ways echoes this ambiguity: in order to make an image that embodies certain aesthetic qualities I desire, I have to heat the bitumen, releasing its toxic fumes into my immediate environment. To protect my own health, I wear a respirator and other protective equipment while working with the bitumen—and yet I still know that some of the toxic material will inevitably leak through my safety gear. While creating petrographs, I am always aware that I have placed myself into an intimate relationship with this toxic and powerful material.

And perhaps so have we all, in our daily lives. The difference is that for most of us, most of the time, we sublimate that relationship so we can maintain a sense of ourselves as autonomous, in control, modern. This is not to say we humans shouldn't strive to imagine ourselves otherwise—indeed, we urgently need to do so—but the ideologies of convenience and empowerment and separateness that underwrite petromodernity are incredibly insidious. For me, the practice of creating petrographs is one way of combating the many forces that aim to anaesthetize us into a false sense of untouchability. The bitumen's physical presence embodies an insistent kind of counter-knowledge. When I smell my bitumen as I am applying it to the photographic plates, or when it sticks to my hands or stains my clothes, there is no way of maintaining that fiction of separation. I have become part bitumen, and it has become part me.

Undoubtedly, this reminder of my petroleum-infused self is uncomfortable, but the bitumen I use for my petrographs does not always conjure associations of toxic complicity. The meaning of the bitumen changes when I move from the lab-like space of my photography studio (structured around containment, consistency, replicability) to the more complex and fertile realm of the field—in this case, the banks of the Athabasca river, where I gather my bitumen samples in their natural environment. I source my material not from the gaping damaged landscape of the tar sands mines nearby, but from the undisturbed boreal spaces that remain on the Athabasca, where the bitumen gradually seeps to the surface in small, thick flows and pools. Rather than viewing this landscape from an aestheticized distance through a camera lens, I must be immersed within it actively, searching for elusive signs in the undergrowth that may point toward small accumulations of tar. When I gather bitumen, I feel like I am relating to petroleum in a way that few people do in the age of petromodernity. During those warm summer afternoons of searching in the forest, the bitumen is not a commodity for me, nor is it a toxic-smelling, sticky mess. Instead it is undeniably a natural substance, one that is entirely and appropriately *in place* there in the boreal forest where it seeps among the layered presence of the other organic materials and beings that have accumulated there: mosses, leaves, pollen, insects. When encountered in its place, it is not an interruption of nature, not a threat to life. In fact, it is often surrounded by some of the richest and most verdant plant life I have encountered in the region.

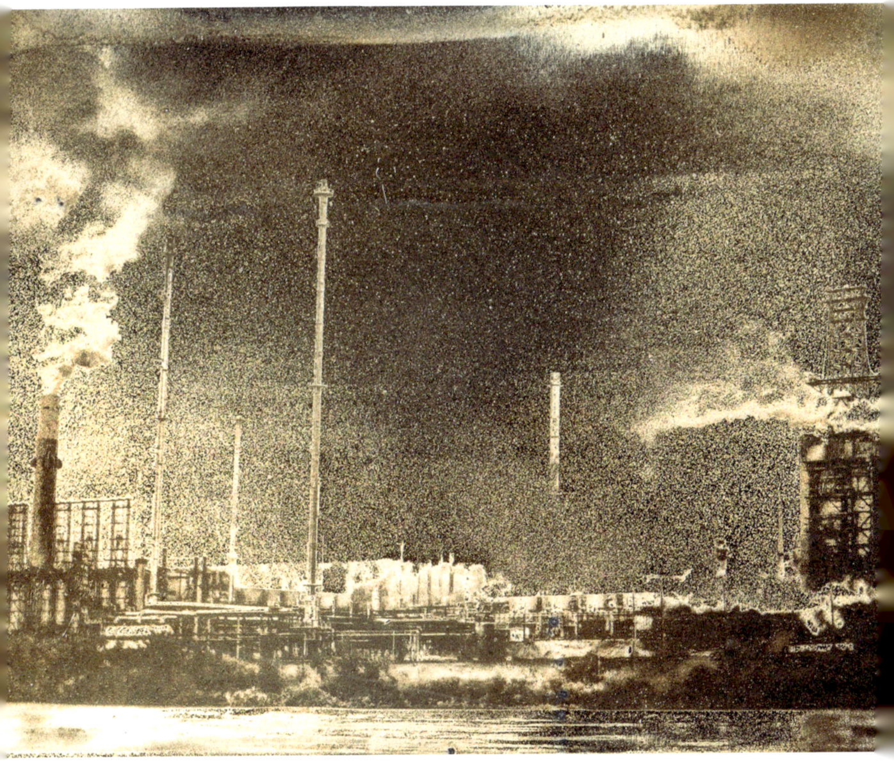

Warren Cariou, *Suncor–Smoke, Steel and Water. V 1 of 2.*, 2014. Athabasca Bitumen and Lavender Oil on Aluminum Plate, 8.5" x 11".

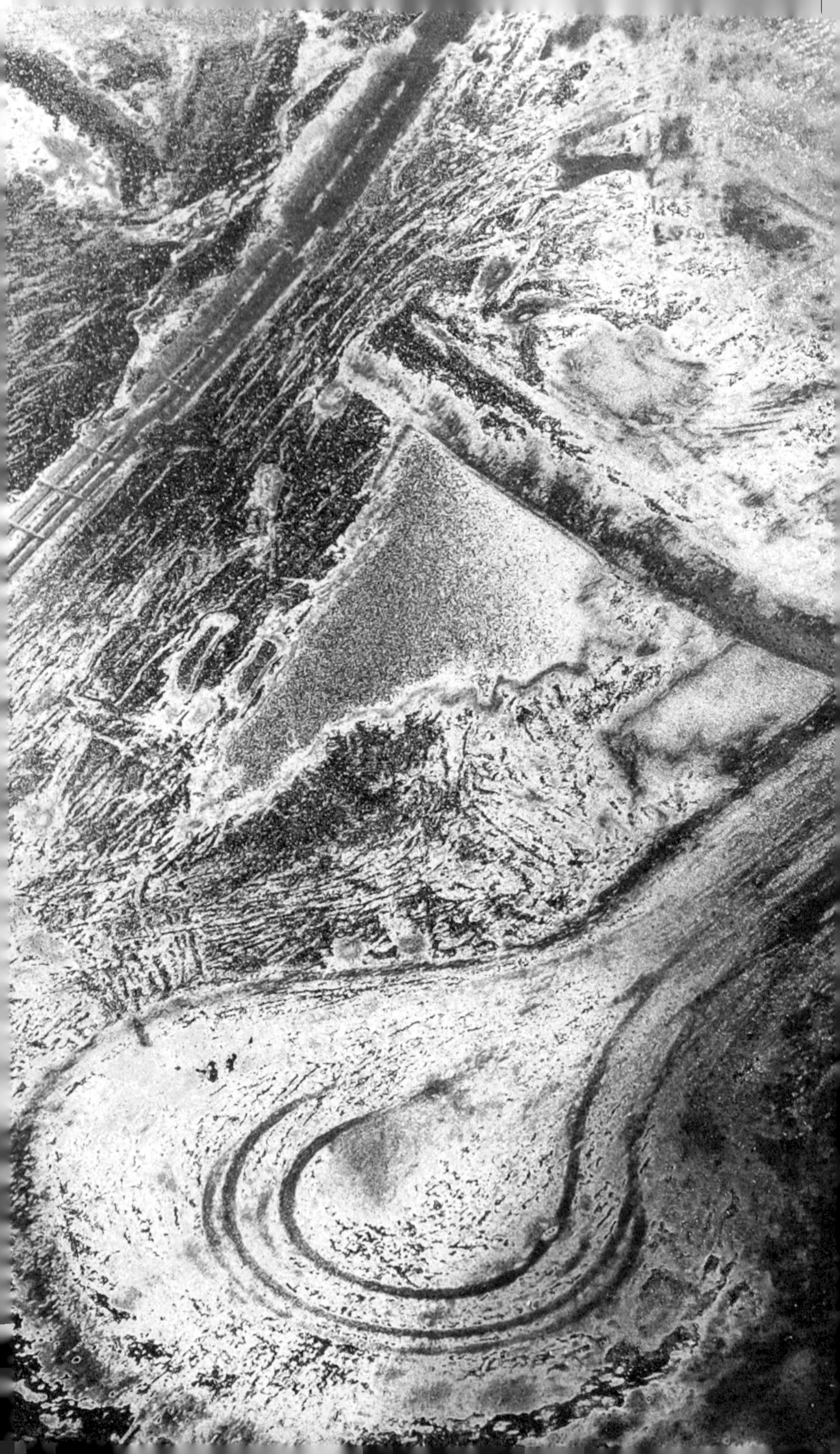

In this context, it is somewhat easier to imagine a different kind of relationship to bitumen, one in which it is not an addictive and fetishized threat to our existence, but instead something like what my Cree and Métis Elders call medicine. My activity of gathering bitumen is similar in some ways to the practices of traditional harvesters who seek out various powerful substances in the land. Indeed, generations ago, my own Métis voyageur ancestors might have gathered bitumen from this same river bank to waterproof their canoes or even to cover and heal open wounds. When I travel in Athabasca territory now, I often think about those ancestors and their engagement in a very different regime of energy, one in which human muscle power and intimate knowledge of the river's surging vitality were the key components. For them and other Indigenous people in the age before the rise of gasoline, bitumen would have been an important medicine that could be used to bring healing or to aid in mobility. But they would also have understood that all medicines can become dangerous if people use them improperly, without the necessary respect and knowledge.

The Indigenous traditions around gathering bitumen in the Athabasca region seem to have been forgotten, or at least I have not been able to find anyone who remembers them. So I have had to make my own way in this practice, knowing that others have gone before me, generations ago. When I find a seep of bitumen where I want to collect a sample, I

Warren Cariou, *Bitumen Strip Mine with Pipeline and Truck Turning Loop. V 1 of 3.* (detail), 2014. Athabasca Bitumen and Lavender Oil on Aluminum Plate, 8.5" x 11".

leave an offering of sweetgrass or tobacco at the site, to give thanks and to acknowledge my ongoing relationship to the place, my responsibility toward it. Then when I bring that bitumen back to my studio and begin the process of turning

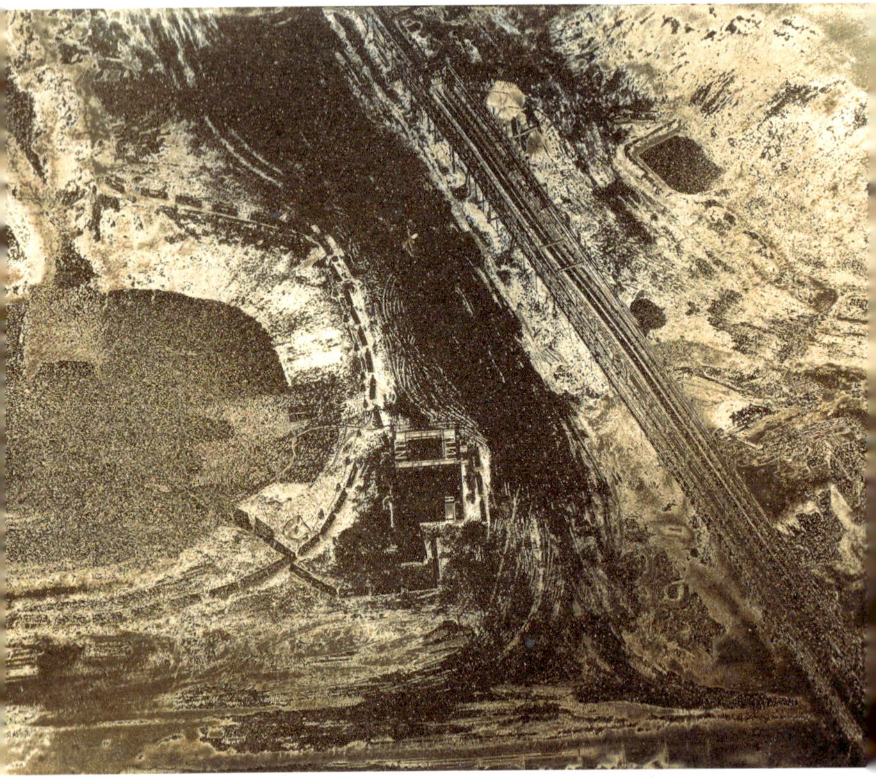

Warren Cariou, *Open Earth with Pipeline. V 1 of 3.*, 2014.
Athabasca Bitumen and Lavender Oil
on Aluminum Plate, 8.5" x 11".

it into a petrograph that will represent what is happening to the land of the Athabasca region, I feel like I am taking another step in that relationship. While I still have to be very careful with the material, taking precautions to protect

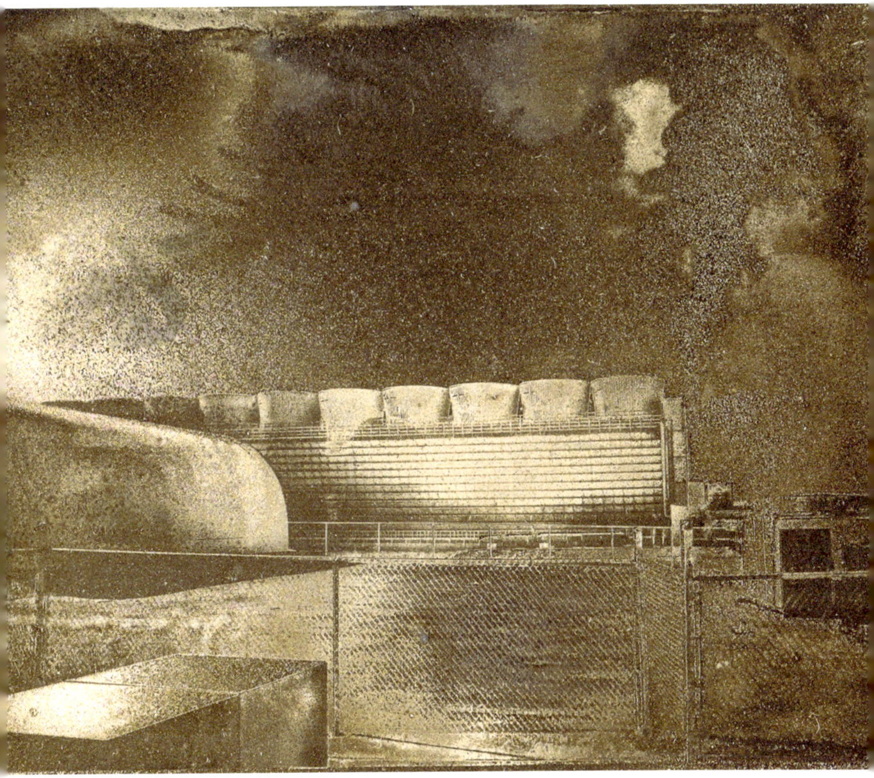

Warren Cariou, *Cooling Towers. V 1 of 2.*, 2014.
Athabasca Bitumen and Lavender Oil
on Aluminum Plate, 8.5" x 11".

myself from its harmful potential, I understand that I am working *with* it rather than against it. Each petrograph is a product of that collaborative relationship that ties me ever more intimately to the land.

7 Protocols for an Ungrid-able Ecology

Kinesthetic Attunements for a More-than-Natural History of a Black Oak Savannah

Natasha Myers

Can art change science? Art/science collaborations could change the ways both artists and scientists think and work, and the questions they ask. Too many of these collaborations, however, leave the sciences intact, as if they are somehow impervious to the provocations of artistic experiment. The most generative collaborations are of course those that succeed in changing the science, altering the ways scientists apprehend the world, getting them to question not only their their experimental techniques and modes of attention, but also the founding logics that animate their practice. This short provocation stands alongside Jennifer Willet's remarkable art/science experiments that push up against the domains of the natural and the unnatural. And here I suggest that in addition to feminist interventions in the sciences, artists can also work to decolonize the sciences. This short essay invites artists to challenge scientists to decolonize their attentions and perceptions. This is important political work: the colonial legacy of the sciences has shaped how and what we know; the sciences render the the world legible in some ways and not others. Following Donna Haraway we need to continue to ask "for whom and at what cost"?[1] What forms of life, what beings and doings, remain illegible and inconceivable to the sciences? What is not seen, said, felt, imagined or known? What cannot be captured by the metrics of a capitalist, expansionist and extractive knowledge system hooked on the lie of its universality and neutrality?

What if it were possible to loosen the grip that the sciences have on our imaginaries about claims to truth? What if we gave up on the lie that science is a rigorous practice of disinterested objectivity; that scientists can enact a "god-trick" to can see "everything from nowhere;" and that facts are "out there" waiting to be discovered?[2] I propose that art/science convergences might be more generative if they began with the assertions that 1) *science is not what we thought it was*, and 2) *it could be otherwise*.

Working as an anthropologist among scientists for the past fifteen years (after training as a molecular biologist and a dancer, and having launched a series of art/science collaborations), I have learned remarkable things about the synesthetic, affective, and embodied nature of scientific practice. The protein modelers I have worked among—those who build and use atomic resolution models of protein molecules to render visible the chemical structures of the stuff of life—have shown me just how thickly thinking and feeling are tangled up with one another in the craft of science.[3] Over the long durée of building protein models, they not only sculpt an atomic resolution model of their molecule on a computer screen, they also tune their kinesthetic imaginations to protein form. Defying what we thought anthropomorphism was, their molecules become wily and lively just as the modelers themselves *become molecular*. Modelers practice a near-shamanic art of shapeshifting, where the expressive, gestural body becomes the most accurate and ready-to-hand proxy for the molecule. In their hands, and through the lively stories they tell with their animated bodies, models and molecules readily morph between the machinic and the animal. These modelers have taught me that model making is an affectively charged and kinesthetically attuned material-semiotic practice that *rends* matter and *renders* meaning.

These modelers' remarkable practices demonstrate that objectivity is situated, embodied, felt, and relational; rigour is more and other than routine, regulated, exacting, undistracted, or detached. They practice science as if their task was to invent techniques to become more fully entangled with their objects and cultivate deep, abiding, careful modes of attention. They practice science as if rigor has more to do with passionate attachment than neutrality; as if robust forms of knowing can be generated from deeply sensory, sensual and near numinous attunements to the excitability of matter. They articulate clearly how their deeply intuitive feelings for molecular facts can be transduced through their own excitable tissues. In the process, they animate their knowledge by literally dancing molecular worlds into being.

These ethnographic encounters reveal that scientists continually disrupt the conventional consensus about what counts as a fact, what counts as data, what counts as objectivity, and what counts as knowledge. Above all, these scientists taught me about the failure of mechanism to fully disenchant the life sciences and their failure to adhere to the mechanistic, functionalist, neo-Darwinian logics they think they are supposed to avow.[4]

The major problem, however, is that it is not only the scientists that think they are supposed to uphold these foundational logics: scientists' publics, institutions, stakeholders, funders, and legislators do too. In a post-truth era of alternative facts, even the most critical thinkers are reviving calls for that mythic form of disembodied scientific objectivity in the hopes that it can generate clean, cold, clear data. Hard facts are at a premium just as scientific research funding and federal agencies in the USA and Canada shut down their research programs.

While to many this seems like no time to be taking aim at the sciences, perhaps this is precisely the time to call for robust modes of knowing that do not reproduce the colonial, militarized, capitalist logics on which science was founded. I want to hold out for forms of inquiry that can decolonize and de-militarize the sciences, for modes of attention that do not reduce land and bodies to alienable commodities and deterministic machines.

The scientists I've worked with showed me that science can be practiced otherwise. There is a remarkable resonance between artistic practice and the forms of inquiry that protein modelers taught me were immanent to producing scientific facts. I want to see art/science convergences build on the embodied, synaesthetic, affective dimensions of science to craft forms of knowing that can hold scientists, governments and industries accountable to the needs of communities fighting, for example, the toxic ecologies of late industrialism and climate change.

Alongside Jennifer Willet, I hold out hope that art might be the best source of dissensus to begin the work of rendering science otherwise. In what follows I document a few of my efforts to ask: Can art practice change how we apprehend nature? Can it change the methods we might use to study ecological phenomena? Can it alter our modes of attention and de-tune our colonial ecological sensorium so that we can cultivate other ways of knowing?

De-Tuning and Re-Attuning the Ecological Sensorium

The lands on which Toronto stands today used to be covered by oak savannahs. An oak savannah is composed of widely spaced oak trees, tall prairie grasses, and wildflowers. This particular composition of vegetation loves to take root in the sandy soils of ancient lake beds. Oak savannahs also thrive on the disruptive force of fire: fires ensure long-term regeneration of the land.

Fire is of course not just a "natural" force; people all over the world use fire to sculpt lands. Oak savannahs depend on people with knowledge of fire and the skills to care for the lands. Toronto's remnant black oak savannahas, including those in High Park, are millennia in-the-making.[5] These lands are the traditional territories of the Wendat, the Haudenosaunee and Anishnaabe Nations. Toronto stands on the lands of the Mississaugas of the Credit River. Indigenous peoples cared for this land with fire for millennia before colonization. Many thousands of Indigenous and Métis peoples live and move through this region today.

Just a fraction of the lands that once thrived as oak savannahs survived settlement. What do the remaining remnants of those oak savannah lands remember from a time before colonization?

Oak savannahs do not survive without people. For over a century settlers grazed their sheep and later their lawn mowers across these lands, keeping the savannah lands open, but suppressing the regeneration of the oaks and the grasslands. Today, Toronto's Urban Forestry team conduct controlled burns annually in an effort to bring back the oak savannahs. And yet, these efforts to save the park's "nature" come with no attention to the rich cultures that

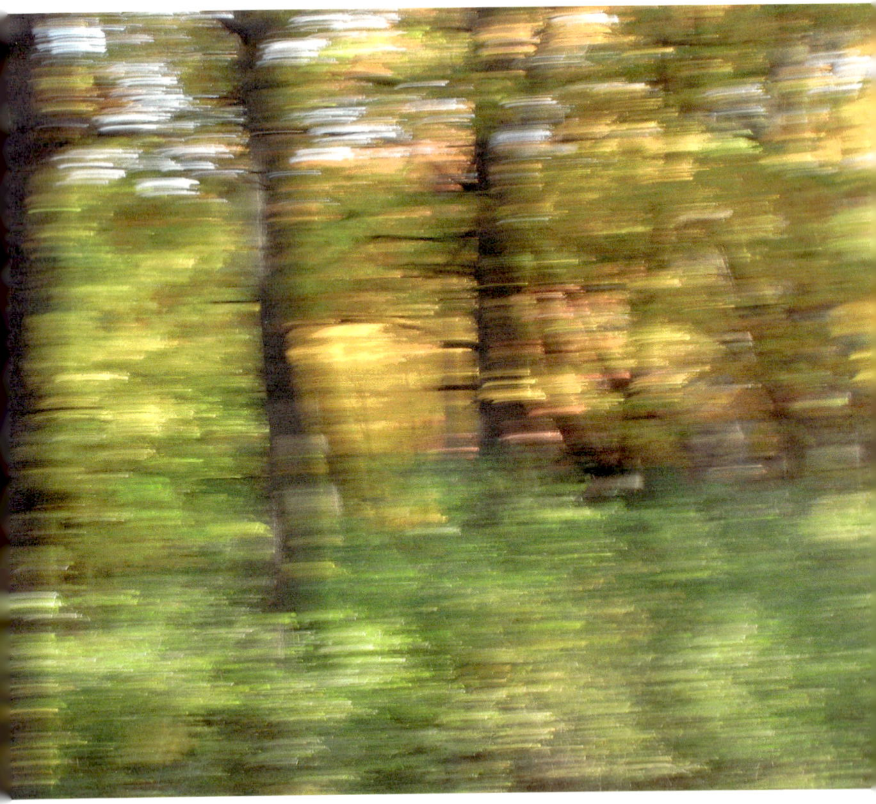

Natasha Myers, *Oak savannah lands*, 2016.
Color photograph.

gave this land its contours and significance. In this sense, restoration efforts participate in an ongoing colonial project that continues to enforce the dispossession of Indigenous peoples from their lands. Can we do ecology otherwise?

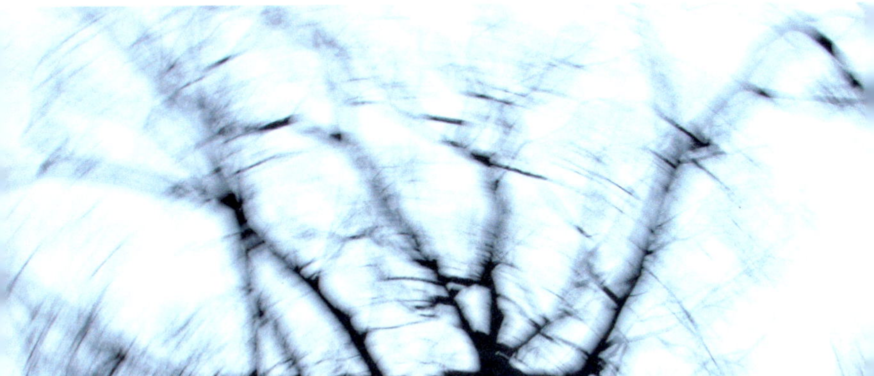

Natasha Myers, *Oak sentience*, 2017.
Color photograph.

Oak savannahs are *naturecultures*;[6] they are sites where people have learned how to *conspire* with the plants.[7] Is it possible to decolonize our vital plant/people conspiracies in order to imagine other possible futures?

Such questions are the focus of *Becoming Sensor in Sentient Worlds*, a research-creation collaboration that invites you to *decolonize your ecological sensorium*.[8] To do this you need to forget your best training: forget what you thought "nature" was; forget how you thought life "worked;" and forget, too, the naturalizing tropes that made you believe that living beings "work" like machines, or that forests perform "ecosystems services," or that "reproduction" and "fitness" are the only valuable and recordable measures of a life.

A collaboration with award-winning dancer and filmmaker Ayelen Liberona and field recordist Allison Cameron, this project aims to make strange the ways that the conventional ecological sciences have not only been deployed to colonize land, but also to colonize our imaginations and how they evacuate all other ways of knowing the living world, most especially those local and Indigenous knowledges that are attuned to the sentience of lands and bodies.[9] *Becoming Sensor* invites you to sensitize yourself to the power moves of a settler colonialism that has rendered more-than-human sentiences so illegible, and so impossible to perceive.

Protocols for an Ungrid-able Ecology

What would change if you knew that the trees were watching you? What do the trees of High Park's oak savannahs know? If you learned how to listen, what stories could they tell? *Becoming Sensor* asks what modes of attention need to be cultivated to pay attention to these lands that have been paying very close attention to all the transformations taking shape around them over millennia.

We take the "environmental monitoring" plots strewn throughout the park as provocations. What would be required to really pay attention to this land? How could our sensing practices do justice to documenting the beings and doings, not only of an oak tree, but also the vast numbers of creatures rooting, weaving, and winding their ways across these lands? *Becoming Sensor* in the savannah demands subtle attunements of our always already *synesthetic* sensoria. It demands cultivating new modes of embodiment, attention, imagination and new ways of telling stories about lands and bodies.

Working at the cusp of art, ecology and anthropology, *Becoming Sensor* aims to *do ecology otherwise*. The images, sounds and energy diagrams we generate in the field are "data" for an *ungrid-able ecology* of this *naturalcultural* happening. Our protocols reinvent ecological modes of attention and data forms by cultivating synaesthetic attunements to the land through forms of *kinesthetic imaging and kinesthetic listening.*

Kinesthetic Imaging

Technique: Hack into your camera to slow down the shutter speed so you can record the energetics of your encounters with the moving bodies of the creatures you are drawing into view. This will allow you to tune in the affectively charged relations taking shape on the land.

Considerations: If traditional nature photography captures living bodies and turns them into objects of aesthetic and

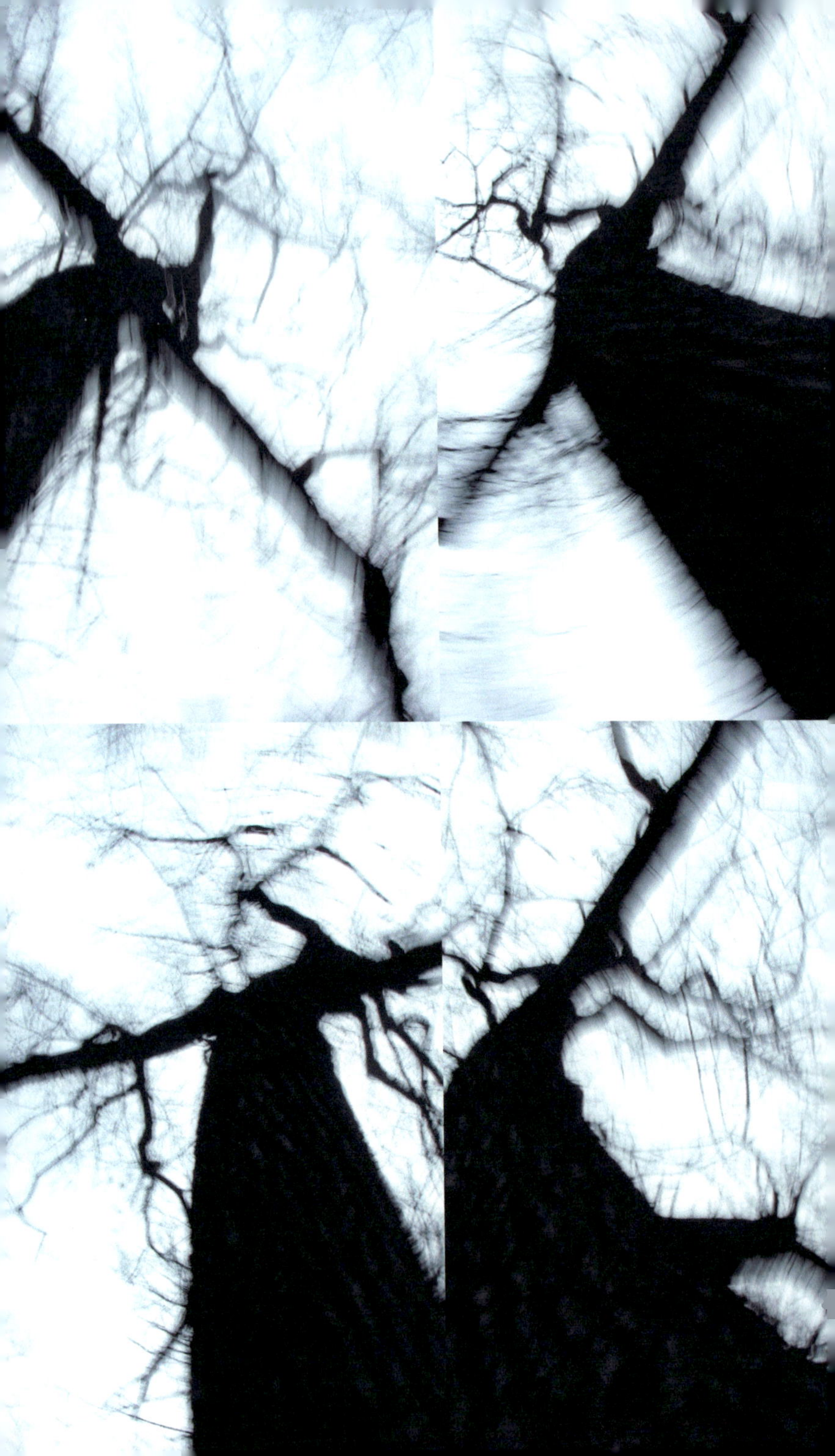

scientific interest, these kinesthetic images gesture to a different kind of account of the living world. These images are attunements. They are generated in the act of *moving with and being moved by* the beings and doings around you. As relational images they document the push and pull between bodies. Rather than a means to capture events or objects, these images make it clear that it is the photographer who is caught: captivated, they are the ones who hitch a ride on what is *becoming* and *coming undone.* The rotting logs, frilled mushrooms, crumbling leaves, ancient sands and greening grasses are not discrete things, they are *happenings* taking shape through deep time and in the ephemeral moments of now, and now, and now. It is the photographer who must learn how to keep pace with these rhythms through her body.

Note from April 2017: Documenting dark ecologies on night walks in the oak savannah: I hold the aperture of my camera open for 13 seconds as I walk the paths, camera held at my chest. Colorful lights from the roadways penetrate deep into the night savannah. Each step, each breath, each shift in my body as I navigate these lands is recorded in the play of light across the image. These kinesthetic images hitch a ride on the mood of the descending darkness in the savannah just as they keep my embodied experience in view. The boundary between urban life and park life is blurred in new ways.

Natasha Myers, *Dances with trees*, 2017.
Color photographs.

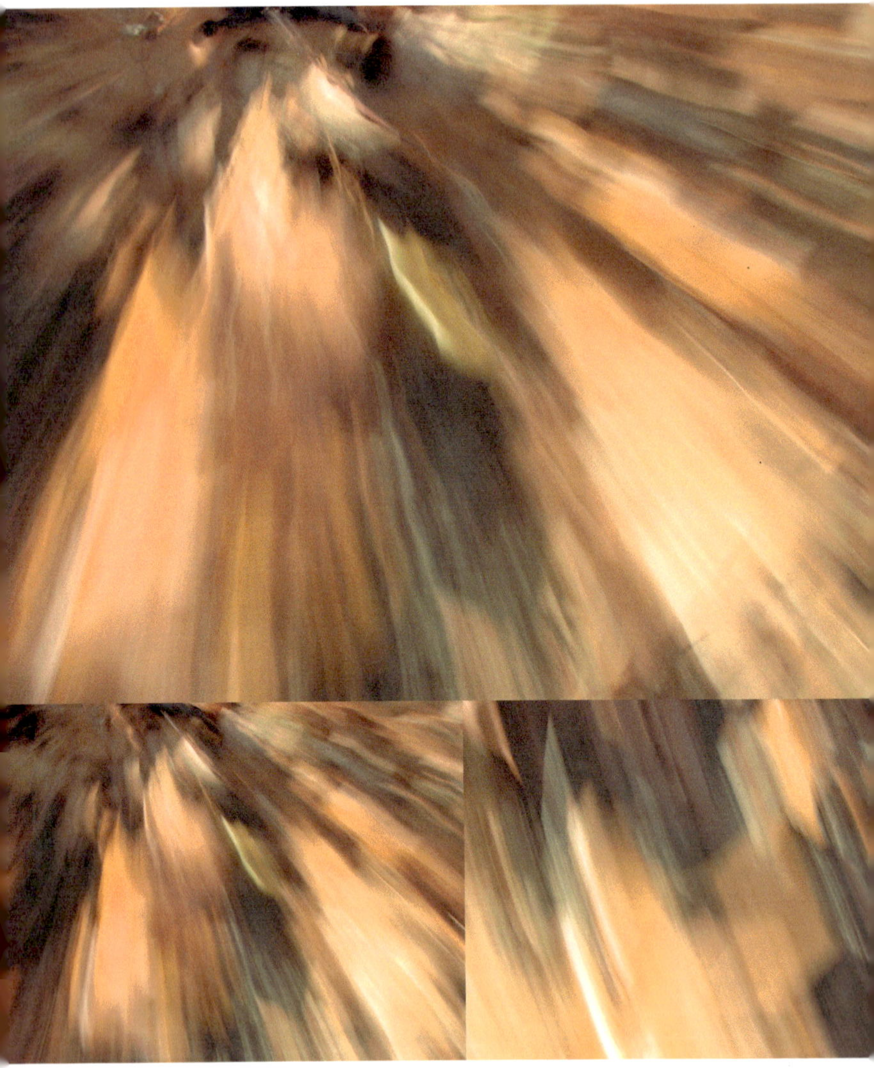

Natasha Myers, *Compositions and decompositions in the savannah*, 2016. Color photographs.

Kinesthetic Listening

Technique: Bring a digital audio recorder with a directional microphone with you into the savannah. Rather than standing still, stay in movement while you are recording. Lean into the sounds to amplify their intensities, speeds, slownesses and their affective charge. Turn and twist your body to feel through the multiple sources and trajectories of sounds. This dancing with sound is a mode of *kinesthetic listening*. Back at home, use sound editing software to speed up and slow down the sounds. What do you hear?

Considerations: Sound palpates space and pulls at time. It is a remarkable tool for documenting ecological relations. Document the vibratory milieu of the savannah by tuning into its deep time, its seasonal cycles, its daily rhythms, its improvised encounters, fleeting moments and disruptive events.

Note from April 2016: The sounds of cars and trucks and planes are never fully muted here in the oak savannah. They just propagate differently. Muffled and modulated by trees and shrubs, birds and squirrels and insects, ravines and slopes, city sounds resonate in a distinct vibratory milieu. *Sounding out the savannah* reveals that there are no boundaries between the rhythms of city life and the lives of the creatures who take root and take flight here. Speeding up and slowing down the recordings reveals otherwise unimaginable worlds and opens up new ways of telling stories. Stomping feet become falling trees, shaking the earth in ways that recall the geological forces that formed this land. Slowing down bird calls reveals other songs, other creatures and voices

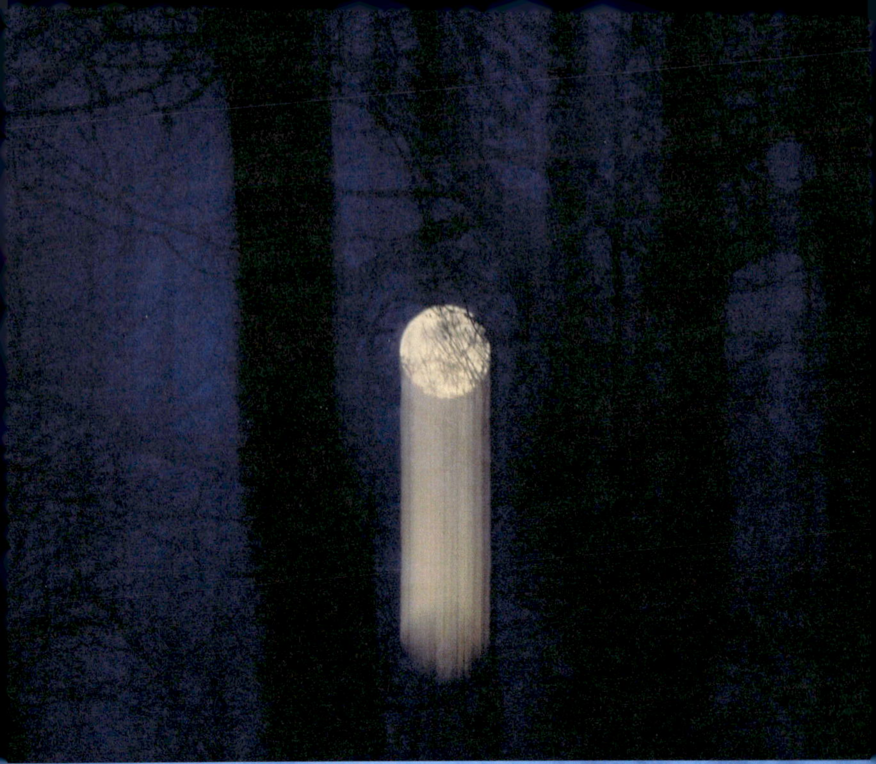

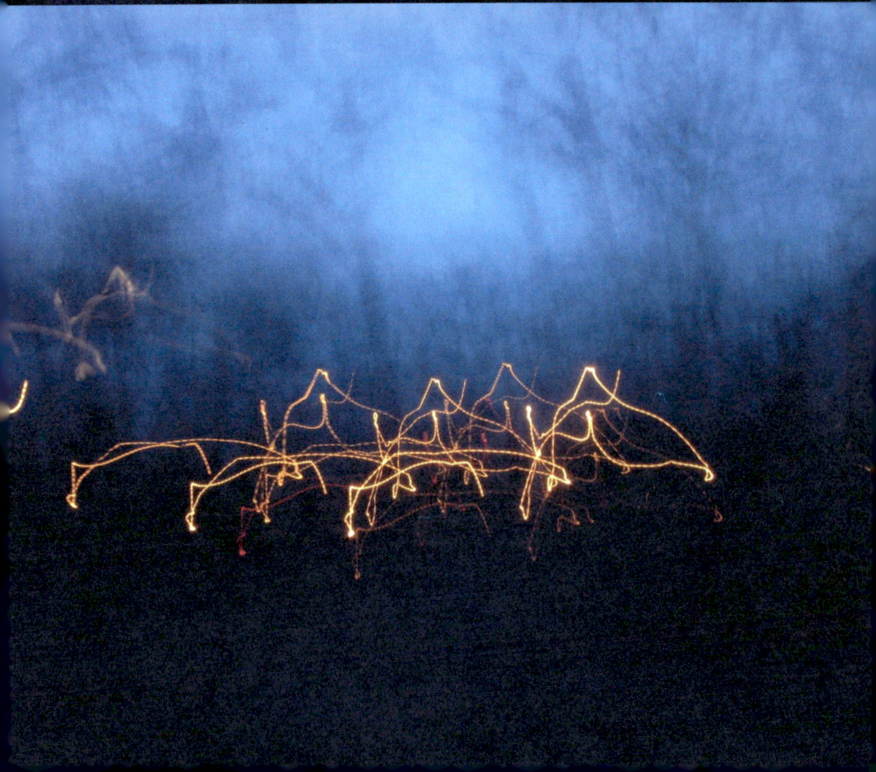

haunting the space. Gulls become coyotes. Traffic becomes rushing, rhythmic waves. Life churns to other rhythms. There is no silence here.[10]

Kinesthetic Sniffing

Premise: To become a nose capable of sniffing out plants' alchemical utterances in an affective ecology one must begin from the assumption that plants are creative and expressive synthetic chemists mattering and modulating the chemical composition of the atmosphere. Plants are alchemists who craft volatile concoctions to excite other plants, as well as animals and insects and people. The aim is to tune in to the significances and sentiments that plants articulate through their volatile chemistries.[11] Mapping these smells is a way to map *involutions* among plants, among plants and insects and among plants and people.

Technique: To do this work you must become a *transducer* of the alchemical utterances shaping this affective ecology. You must let these scents excite your tissues in order to document your situated knowledge of this alchemical ecology. And you must learn to transduce chemical excitations through your tissues and record these as energy diagrams.

TOP: Natasha Myers, *Full moon rising over the night savannah*, 2017. Color photograph.

BOTTOM: Natasha Myers, *Dark ecology*, 2017. Color photograph.

Step 1: Sniffing out the Savannah

a. Select individual flowers or clumps of flowers. Lean right in to the flower to take in its full bouquet. Pull away. Lean back in for a second sniff, and then a third. Let the range of smells move through your tissues. Feel how the scent has form, speed, height, depth and weight.

b. Transduce each sniff as an energy diagram, letting the pencil in your hand loosely track the energies of this encounter.

c. Note plant species, the time of day, the date and time of the last rain, the location and describe the composition of neighboring plants.

d. Repeat at different times of day and in different points in the season.

Step 2: Develop an ambulatory mapping of this smellscape over time.

Considerations: Pay attention to atmospherics and temporalities: it seems as though evenings are a time when the scents linger, when they are no longer getting burned off by the sun or evaporated in the heat. When the sun's rays lengthen and the scented air cools the smells seems to drop down and gather in the troughs of the savannah's undulating landscape. Note where the smells mingle, and where the contours of the land shapes the ways the smells gather.

NOTE from August 1, 2015: Be careful: the flowers are potent. Do not work with more than one species at a time. I got totally high after working with sweet clover, wild bergamot, Queen Anne's lace and goldenrod in one session.

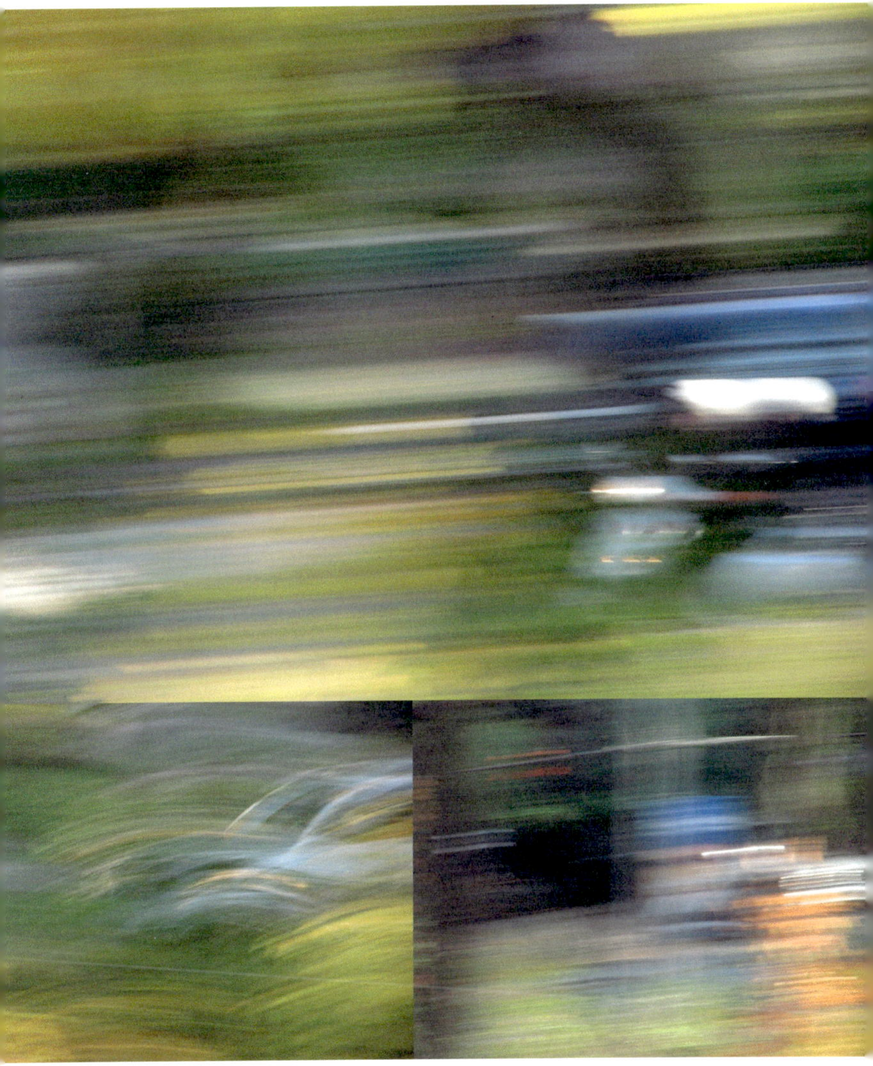

Natasha Myers, *Grass, leaves, glass, metal, concrete, and petrochemicals: a more-than-natural history of an oak savannah*, 2016. Color photographs.

Coda

These protocols are not intended to re-colonize these lands with yet more settler colonial stories. Nor do they appropriate Indigenous knowledge or practice. Rather, these techniques offer an example of some ways to break the consensus of an ecology that is indebted to capitalist

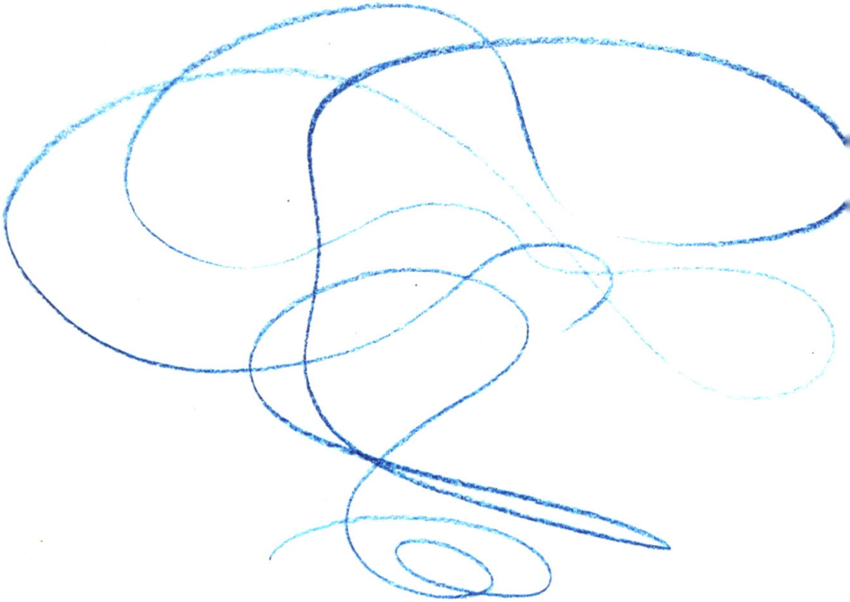

Natasha Myers, *Energy diagram of Queen Anne's Lace*, 2015. Pencil drawing.

and colonial logics. These art practices refuse the flattening universalism and mechanism of the sciences. By altering modes of attention and perception they push up against the forces of a scientific rationalism that disavows nonhuman sentience and commodifies nature as resource. In so doing, these techniques aim to expand the discursive field in which stories about lands and bodies can be told. Consider these attunements as one way that settlers can ally themselves with the remarkable work of Indigenous activists and scholars in the name of decolonization.[12]

You too can *do ecology otherwise*. Try your hand at inventing techniques that disrupt the colonial ecological sensorium. Remember, though, that an *ungrid-able ecology* is not bound to the moral economies of mechanism or the energetic efficiencies of neo-Darwinian survival stories. And as you experiment with techniques, modes of attention and inquiry, and get interested and involved in the *naturalcultural* happenings around you, be sure to remember to resist the compulsion for legibility, quantification and grid-like mappings. What modes of attention does the land you dwell on demand? What other worlds can you perceive? What other futures can you conjure?

Notes

1 Donna Haraway, *Simians, Cyborgs, and Women: The Reinvention of Nature* (New York: Routledge, 1991).
2 Ibid.
3 Natasha Myers, *Rendering Life Molecular: Models, Modelers, and Excitable Matter* (Durham: Duke University Press, 2015).
4 Ibid.

5 See J. L. Riley, *The Once and Future Great Lakes Country: An Ecological History* (Montreal: McGill-Queen's Press, 2013) and J. Johnson, *Pathways to the Eighth Fire: Indigenous Knowledge and Storytelling in Toronto*, Doctoral Dissertation, Graduate Program in Communication and Culture, (Toronto: York University, 2015).

6 A. L. Tsing, *The Mushroom at the End of the World: On the Possibility of Life in Capitalist Ruins* (Princeton: Princeton University Press, 2015).

7 See Tim Choy, "Breathers Conspire—On Drawing Breath Together," Paper presented at the Annual Meeting for the Society for Social Studies of Science (Barcelona Spain, September 2016).

8 Ayelen Liberona and Natasha Myers, with A. Cameron, *Becoming Sensor in Sentient Worlds*: An immersive, synesthetic installation, 2017. http://becomingsensor.com

9 See Leanne Betasamosake Simpson, *Dancing on Our Turtle's Back: Stories of Nishnaabeg Re-Creation, Resurgence, and a New Emergence* (Winnipeg: Arbeiter Ring Pub., 2011) and Leanne Betasamosake Simpson, "Land as Pedagogy: Nishnaabeg Intelligence and Rebellious Transformation," *Decolonization: Indigeneity, Education & Society*, 3.3 (2014), 1-25.

10 Ayelen Liberona and Natasha Myers, *Becoming Sensor in an Oak Savannah*, 2016. https://www.youtube.com/watch?v=J02XTtDmCrU. See also Natasha Myers, "Becoming Sensor in Sentient Worlds: A More-than-natural History of a Black Oak Savannah," in *Toward an Artful Anthropology: Between Matter and Method, eds.* G. Bakke and M. Peterson (London: Bloomsbury Academic, in press) and Natasha Myers, "Ungrid-able Ecologies: Decolonizing the Ecological Sensorium in a 10,000 year-old NaturalCultural Happening," *Catalyst: Feminism, Theory, Technoscience*, special issue "Science out of Feminism," forthcoming.

11 see Carla Hustak and Natasha Myers, "Involutionary Momentum: Affective Ecologies and the Sciences of Plant/Insect Encounters," *Differences 23*.3 (2012), 74–118.

12 See for example: Audra Simpson, *Mohawk Interruptus: Political Life Across the Borders of Settler States* (Durham: Duke University Press, 2014); Zoe Todd, "From a Fishy Place: Examining Canadian State Law Applied in the Daniels Decision from the Perspective of Métis Legal Orders," *TOPIA: Canadian Journal of Cultural Studies* 36 (2016), 43-57; Eve Tuck and K. Wayne Yang, "Decolonization Is Not a Metaphor," *Decolonization: Indigeneity, Education & Society* 1 (2012), 1–40.

8 Good Housekeeping

George Gessert

Years ago at a crafts market in Tlaquepaque, Eric bought for a few pesos one of those little wooden boats crammed with pairs of tiny animals. "Isn't it wonderful?" he said to Sally. The colors, reds and yellows mostly, seemed to radiate good cheer; perfect, Eric thought, for Western Oregon where forty days and nights of rain are common, or at least they were before the weather became erratic.

Eric put his find high up on a bookshelf in their living room. At the time he believed that the ark was about good stewardship and love of life, animals in particular. True, it was also about inconvenience: Noah had to listen to God, defy convention and build the ark, none of which could have been easy. But in the end everything turned out fine: sinners were punished, the world was washed clean and Noah's family repopulated Earth. Imagine repopulating Earth! The story was so agreeable that he soon forgot about the ark.

Then there came the diagnosis. Cancer casts a new light on things, including tchotchkes. He couldn't help but notice those reds and yellows: they were bright, too bright, intense to the point of alarming. Venomous creatures display them. He reread the story of the Flood and found it as shocking as *Oedipus Rex*. Yahweh drowns not only sinners, but their children, including babies. People had lots of children in those days. Did He drown them to punish their parents?

Was it collateral damage? Or maybe it was preemptive: if those babies grew up, each of them was bound to sin.

Then again, maybe sin wasn't the point. Maybe we are punished for being born—that's what the Greeks thought. It would certainly explain the animals. They are incapable of sin, yet Yahweh destroyed virtually all of them. Even the remnant, those pathetic pairs driven, dragged or carted onto the ark, were doomed to environmental havoc and genetic bottlenecks once the waters receded. Only a handful of scavengers—rats, dogs, flies—had much to look forward to. For them there would have been carrion to go around, just as there was for our shrew-like ancestors after the asteroid wiped out the dinosaurs.

As for plants, *Genesis* all but ignores them. The operative assumption was that plants don't need an ark to survive. However, as any gardener knows—and Eric was a gardener—few land plants, including seeds, can endure for long under water. Olives need well-drained soil and would have been among the first to go. The dove must have returned to the ark with some other bit of greenery, from something tolerant of prolonged immersion, like cattails or reed canary grass.

Cancer had filled Eric with rage. On the surface nothing had changed. He went about daily life as usual, but all the while he was waiting for an opportunity to destroy—something. It made him feel closer to the gods. They revel in destruction. Eric had to admit that at times he wanted the human race gone, most of it anyway, but he drew the line at mass extinction. True, he contributed to it himself, not least by undergoing extremely expensive, energy-consuming treatments for a disease which in the end was almost certain to kill him, but he never willed extinguishing the majority of species. He was determined to live, which is the oldest, lamest excuse for wanton killing, but not the same as delivering the command. No, not the same: he decided that the Yahweh of *Genesis* was outright mad, that is, psychopathic, not merely angry.

Eric built a fire in the living room fireplace and tossed in the ark. It blazed like an opal, green, blue, and rose. Was it something in the pigments? He gazed at the exquisite flames, and for a moment was a child again, sitting by a campfire listening to stories.

He watched the flames awhile longer, then grasped the ark with tongs and carried it to the kitchen sink, where he doused the flames beneath a tap. Steam rose up, while a leopard's spots, miraculously spared incineration, flowed down the drain. Tempera, he thought. He set the oven at 250 and slipped the charred hulk inside on a cookie sheet. He had a peanut butter sandwich and a glass of orange juice, then removed the ark and let it cool. Back up on the bookshelf it looked like some blackened lump one might come across in a rubble-strewn vacant lot or an old-growth forest.

That evening when Sally returned she smelled smoke. Eric told her what he had done.

She laughed. "I thought you liked it. I never did—no, not even in Mexico. I told you it was garish—yes, I did tell you—but you didn't listen." She gazed at the object on the bookshelf. "It looks much better now. But you should have let me help. We could have had a romantic fire."

9 Hi

George Gessert

Waiting rooms can be a bit depressing, but I enjoy the magazines, especially *Smithsonians* and *National Geographics* with reports on the latest fossil finds. New dinosaurs always lift my spirits. I like to linger over pictures of strange, prehistoric creatures, giant sea scorpions and cephalopods with sutures as intricate as lace. Sabertooths and quaggas are marvels of creation, and once, in a radiologist's waiting room, I saw an Ordovician dragonfly. It was the size of a raven and sailed through a forest of horsetails as rigid as totem poles, some upright, others leaning at precarious angles over black, still water. The hyper-realistic rendition seemed hallucinated, and transformed the waiting room into a natural history diorama where I and other patients gazed at magazines while a meteor arced across the sky.

I also like to browse *People*, *US*, and the glitziest of them all, *In Touch*. Only a few waiting rooms carry it, even though no glossy more reliably distracts from lab reports—at least that was how it always worked for me until the afternoon I had to wait for an hour to see Dr. Compton. If I had not been bored, I doubt that I would have read the article—ordinarily I don't read *In Touch* I just look at pictures—but the title grabbed my attention: "Stage 4 Triumph Over Insurmountable Odds."

She lived with her husband, two young children, and a golden lab named Sam in Corvallis, Oregon. Life was hardly glamorous or Jurassic, but it was good, very good, until the diagnosis. She had pancreatic cancer and "maybe" six months to live. After that things went from bad to worse. The day she began chemotherapy her husband fell from a ladder while cleaning out a gutter. For weeks he lay in a coma. Overwhelmed, she forgot to feed Sam, who ran off and had a fatal encounter with a truck. Racked with guilt, she took to sleeping pills. She thought of suicide, but one morning, as she was pouring herself a finger of gin, she caught sight of her children playing. Their curls, their laughter—she knew then that she had to fight, if not for herself, for them.

Among her weapons was guided meditation. After mastering certain techniques, she made contact with her cancer. The article did not say exactly what took place, but following her encounter she was cured. Doctors could not explain it. At the end of the article her husband had also recovered and the children were playing with a new puppy.

I told myself that obviously her cure had been due to the chemotherapy, which the article mentioned only in passing, but a spore had been planted. It hatched and ramified in the moist darkness of my subconscious until I too felt the imperative: make contact with your cancer!

And so early one afternoon I closed my eyes, focused on my breathing, and relaxed. Nothing much happened except that I grew drowsy. The next morning I tried again, this time well before lunch, but still nothing. Determined, I consulted my iPhone. It told me how to visualize a journey through the circulatory system in search of cancer cells, along with "all-important tumor infiltrating leukocytes and natural killer cells." As soon as contact was established, "natural healing" would begin.

I followed instructions to the letter. I turned off all electronic devices in my apartment, including the iPhone, then chose a quiet spot, settled on a cushion on the floor, and with my eyes closed, breathed in and out, deep and slow. I cleared my mind and imagined travelling through my veins—or were they arteries? Either way, it was like rafting, except that I was entirely immersed. Somehow I could breathe, and also see even though there was no visible source of illumination. I was surrounded by platelets and white blood cells, some with granular surfaces, others with tiny tentacles. Here and there I encountered what looked like lumpy, stained pillows, cholesterol I presume because when I accidentally bumped into one it stuck to my elbow. Long murky stretches alternated with capillaries and rapids. I was buffeted by my heart, almost trapped in my liver, and revitalized while passing through my lungs, but I did not find a speck of cancer. Nor did I encounter natural killer cells—not that I would have recognized one if it bit me in the leg. I should have had a laminated card like those that snorkelers use to identify sea creatures, but even a comprehensive guide book may not have helped. Like terrorists, the cells stayed hidden in the seething crowd. The exercise was at once soporific and exhausting, and I did not try again.

I had all but forgotten my nonadventure when late one evening, as I was brushing my teeth, I heard a little voice.

"Hi!" it said. It came from somewhere deep inside, high-pitched and perky. "I'm Cancer."

"Cancer!" I exclaimed. I laid down my toothbrush and wiped a fleck of foam from the corner of my mouth. "I've been trying to contact you. We need to talk."

"Yes we do!" it said. "You've been, um, let's just say unfriendly. That surgery was nasty. Really nasty! Then radiation, and now Lupron—what's this all about?"

"Look," I said, "I'm not doing this for fun. I just want to live."

"Well, that's all I want too."

"But if you keep on growing, both of us will die."

"Don't be ridiculous," it retorted. "All I am is a little lump. How could anything as puny as me kill you? Think about it: I haven't even invented the wheel. It's those doctors, isn't it? They make me out to be a monster."

At first I thought Cancer was being facetious, but as it rambled on about how friendly and innocent it was, I began to realize it was quite sincere. I had pictured cancer as vicious and aggressive, cunning, hyena-like, or maybe some sort of alien blob. The thought that it was just a bunch of cells having a party dismayed me. I'd gone through surgery, radiation, Lupron and sleepless nights because of *that*?

I screwed the cap back on the toothpaste and when I spoke, my voice quavered with adrenalin. "I want you out. Increase and multiply somewhere else. Leave. Go away!" I repeated "go away" several times as I stormed out of the bathroom into the bedroom.

Shaking, I took several deep breaths. You can't just walk away from cancer, I told myself. And anger won't accomplish a thing. A heartfelt appeal would be better. On my travels through my circulatory system I'd thought about the woman in *In Touch*. She must have told Cancer that she loved her children, and, through tears, waxed eloquent about how much they needed her. I'd decided that if I made contact with Cancer, I'd build a case along those lines, but more expansive to include all sorts of things I love. I'd appeal to cancer's better side. I'd mention friends, redwoods, *Scott and Bailey*, dim sum and "A Supposedly Fun Thing I'll Never Do Again." Also it wouldn't hurt to draw attention to my kindness and generosity. I could mention how I tip twenty percent (unless service is poor) and help Mrs. Jakoubek with her shopping when her gout flares up. But in the heat of the moment all I could think of was my yellow tabby. “Bandersnatch is twelve years old and has colitis,” I stammered. “What will become of him if I die?”

Cancer sighed. “Twelve years old! His fur is cracked, but his temperature is 101, lovely for the lap. A radiant kitty, no doubt about it. But we need to focus. Take a minute, look at this body, how vast, magnificent. It overflows with riches. Blood, muscle, bones.” Cancer made a slurping sound. “Organ after shining organ, and everywhere symphonies of self-regeneration. So why so uptight? Maybe you just don’t want to share? Well, for your information you don’t own this body. I was born and raised here. It’s my home and I wouldn’t leave if I could.”

I sat down on the bed as Cancer nattered on. Somewhere far above an airplane rumbled, spewing greenhouse gases. It occurred to me then that that if Cancer kept on talking I might have difficulty getting to sleep.

As if it read my thoughts, Cancer launched into a rant, beginning with how I deferred to doctors like someone with Stockholm syndrome. In exhaustive detail, Cancer catalogued my lack of “mellowness,” then attacked my diet. I should eat more hamburger, potato chips, have an eclair. Suddenly Cancer paused. When it resumed its voice was gooey, “But I really want us to be friends.” My life, it wanted me to know, was an “endless smorgasbord” and “the best thing that ever happened.” A sarcastic reply was in order, but in the instant before I could deliver, Cancer whispered, “I love you.”

“You’re my daddy,” it murmured, then burst into tears. “You brought me into this world and now you’re trying to kill me—me, your baby, your little cloud of souls.”

I felt sick.

"But what's done is done." Again Cancer's tone shifted, this time growing conciliatory. "We've been through so much together, you and I. Key Largo, I was there, just a chubby cheeked mutation, a little spot of a thing, adorable, if I do say so myself. You were clueless. You still are but I always hope—for you, for everything. Just think: by the Gulf Stream waters I could have gone the way but the very afternoon we returned to California you booked a flight to Sydney. Karen—wasn't that her name? I'll never forget the cosmic rays at 34,000 feet. Mother's milk. I don't think I've ever thanked you. Better late than never, so: thank you. My bits and pieces began to grow and mutate and divide some more. We really have had fun."

Nauseous, I fled to the kitchen and dialed Dr. Compton, but all I got was the answering service. In the kitchen window my reflection floated over night. 11:42. "Never mind," I told the woman, and sat down by the table, phone in hand. A vase holding a wilted alstromeria lay on its side. Bandersnatch must have knocked it over. I had tried to train him not to walk on the table, but it was futile.

"I'm entrepreneurial," Cancer burbled on. "I'm freer than you. I'm in touch with my inner child. I have what it takes. I have what it takes to take, but hey! That's life."

When my health was good, I'd enjoyed cynicism and clichés. Shallowness, carefully cultivated, affords endless opportunities for evasion, but that was over now. I kept silent.

"The basic difference between you and me is that I'm an optimist and you're a pessimist. We make our own realities, too bad for you, but let's just say you're right and my remarkable fecundity, voracious some have even called it, actually does begin to affect this body. That's where invention comes in. In with the new, out with the old. Creative destruction! The world is boundless! Everything will be better, you'll see. We'll evolve, colonize other planets, download our minds. Immortality—why not? The party's just begun. And don't forget, I'm human. Everything I am I got from you."

"OK," I said. "Here's what I can do. You go into remission and we can coexist. But you'll have to be quiet. Quiet. Can you do that? For both of us?"

Cancer burst into gasping, high-pitched laughter. "Let me run this by my billion cells: Who wants to be a monk?" The voice grew harsh. "You're projecting. All these treatments have turned you into a eunuch. Now you want the same for me."

"Consider my offer," I said. "Take a little time, but not too much."

For a moment the room was quiet. When Cancer spoke again its voice was plaintive and so sad that at first I missed the undercurrent of petulance. "You really don't like me, do you? Nobody likes me. Is it because of my name? Cancer: that was never me. I'm a happy-go-lucky guy."

It began to sing:

"Home, home on the range,
Where the deer and the antelope play."

"From now on call me ... Pioneer." It drew out the word, the sound fading into vistas of grassy plains and majestic, snow-capped peaks. "I want love, love and golden statues. On the capitol dome." Cancer's vision materialized for a moment and hovered like a hologram just above the kitchen table, then faded away. "You're the crabby one." Its voice was growing fainter. "You be Cancer."

It was my turn to laugh. "Sometimes I am," I said. I thought of my carbon footprint. "But I'm trying not to be. And now I need to say goodnight. If you have more you want to say, you can take it up with leukocytes."

"War talk. Save it for fundraisers." The words echoed, as if off distant canyon walls.

When its voice resumed, I could barely make it out. "Fights and battles: boilerplate for obituaries." The last word was so faint it could have been bestiaries.

After that all I heard was a passing car and far-off giggling, like water in a drainpipe. I began to hope, but suddenly Cancer's voice resumed as clear as when it first spoke, "Baby needs new shoes."

Every being has its reasons and excuses, but none are more ordinary than Cancer's. I held my breath and waited. A breeze lifted kitchen curtains and stirred the sassafras outside. I listened to the leaves until the ship's clock on the mantel chimed the half hour, then I rose and switched off lights. When I settled into bed Bandersnatch was ready. Purring, he curled up beside my knees.

10 The Phenomenology of (Non)Habitual Spaces for the Bioarts

Melentie Pandilovski

I have been aware of Jennifer Willet's work since the early days of *BIOTEKNICA*, as well as when she took part in a residency with SymbioticA in Perth, with Oron Catts and Ionat Zurr as part of the 2004 Biennale of Electronic Art Perth (BEAP).[1] At this time she also visited me at the Experimental Art Foundation in Adelaide where I was curating one of the *Art of the Biotech Era* projects.[2] Fast-forward to 2011 when I relocated to Canada, and an exchange of FaceBook messages led me to join Jennifer Willet and twenty other artists, theorists, and scientists for *BioARTCAMP* at the Banff Centre during the summer of 2011. Participants explored intersections between science and art in a fully functional biological science laboratory in the Canadian Rockies. The site was built on the premises of a former internment camp (where Ukrainians were held during WWI) and included a very engaging open house where the public could interact with the artists and scientists. A film crew was also present, consisting of Jeanette Groenendaal and Zoot Derks who were filming documentaries on Adam Zaretsky and the *BioARTCAMP*.

The complexities of establishing an improvised but fully functional biological science lab in the Canadian Rockies in many ways invites a phenomenological consideration of spatiality, giving a deeper meaning to Merleau-Ponty's notion of *habitual space* as a form of space that refers neither to *explicit comprehension* nor *blind reflex*. *BioARTCAMP* as a public space included several forms of art-science rituals, which in turn became a setting for the emergence of

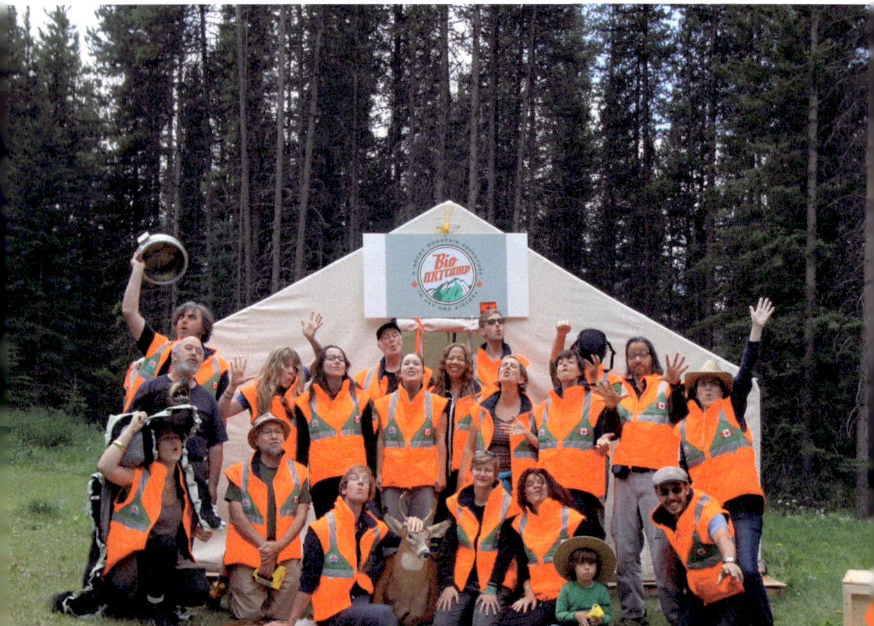

Jennifer Willet, *BioARTCAMP participants*,
Banff, Canada, 2011. Color photograph.

the habitual. If one were to paraphrase Merleau-Ponty one could suggest that *habitual knowledge*, thus acquired, makes knowing and doing work with biotechnology a process that is somewhat intuitive, and certainly one that generates enthusiasm for the creation of *habitual knowledge*. Willet herself looks at the experience with bioart as being transformative, away from the realm of representation, towards the real manipulation of life for aesthetic ends. And indeed, biotechnology never ceases to amaze, through its complexity, radically reconstructing relations between politics and nature, and allowing for a re-assessment of how we look at life today. One cannot help but be astonished and aware of the magnificence of Banff National Park, exemplified by Willet's vivid description of the natural beauties of Banff and the surrounding mountains as a "giant petri-dish!" Only someone fully dedicated to bioart could come to a description like that.

I thoroughly enjoyed the numerous interactions among participants (artists, scientists, curators, theorists, filmmakers), as well as the discussions about bioart, our changing world and its technological, scientific and biological spheres. Needless to say the discussions also included aspects of cultural, artistic, environmental, and ethical thought. My instinctive push was to add Phenomenology, Consciousness, Bio-economy and Biopolitics into the mix as well and act as an *agent-provocateur* with the idea of expanding discussions of what I term the "Formation of the Biopolitical Apparatus." I maintain that it is impossible to speak about biotechnology without speaking about bio-capitalism, as bio-capitalism is the driving force behind the integration of biotechnology into society. The stakes are high. Historically, technological revolutions have always resulted in an alteration of the political, social, and economic spectrums of society. We find ourselves as a society facing radical changes in power relationships in local and international domains. In fact, this biopolitical shift registers at both the economic (bio-capitalist) and cultural (bioculture, bioart) levels. This shift, generated by the Biotech Revolution, configures the biological as political and economic. The biological-as-political includes notions of human rights, the changing and increasingly toxic environment and bioterrorism. The biological-as-economic sees bio-capitalism as the latest stage of Capital's development, but also discloses a certain negation of profit-oriented values and the necessity of growth, thereby holding an ambivalent ethical position regarding capitalist production values. As the consequences of the Biotech Revolution become apparent in the political and economic spectra of society, so too do contemporary biopolitical discourses intersect with the re-contextualization of relations between state apparatuses, scientific protocols and cultural systems. These relations coalesce in the construction of a global biopolitical apparatus, encompassing new vectors of power with regard to social, political, economic, and administrative mechanisms, as well as knowledge structures which have the capacity to create, maintain, or destroy contemporary society. Biotechnology thus enables a certain neo-politicization by putting into motion control mechanisms based on a coding system, altering the dynamics between the state and the individual, and resulting in an increasingly programmable and disciplined society.

It has been clear to me for some time already that as bioartists expand their practice, it becomes very difficult for art critics and cultural theorists to fully understand the new sorts of artworks they produce, let alone find consensus or common ground. How indeed can we define a work that merges genetics, art, and information technologies, augmented and accented through their interactions? Although bioart has been around for three or four decades, it wasn't until 2013 in Krakow that the International Congress of Aesthetics deemed bioart a worthy theme of discussion. Even then, the discussions mostly revolved around Eduardo Kac's most famous projects *Genesis* (1999)[3] and *Alba* (2000),[4] as the esteemed Congress had basically just discovered them. This is not surprising, as in reality it is difficult to decipher what is going on in such works, let alone imagine the full impact of what is to come. The development of biotechnology seems to have the potential to cause a tectonic shift in our culture. Generally speaking, cultural change is accompanied by, even caused by, changes in consciousness. In this case we are witnessing a process of changing our critical view point—away from traditional understandings of civilization and towards an intersection between the engineered and the biological.

In order to interpret the processes surrounding biotechnology and bio-culture we need a new tool. It seems most appropriate to me that in order to find such a tool, we should turn to Phenomenology, referred to by its founder Edmund Husserl as *the science of consciousness.* My phenomenological inquiries are aimed at researching the possible nodes where changes in consciousness cause the replacement of one culture with another, in this case researching the initial phases of the establishment of a Global Biopolitical Apparatus. In this, I raise questions regarding the influences of biotechnology on the structure of consciousness and the relation of consciousness to the phenomenal, taking the lead from Husserl who declared Phenomenology to be the study of the structures of consciousness that enable it to refer to objects outside itself. This involves various methods of reflection, introduced in order to develop Phenomenology into a discipline that

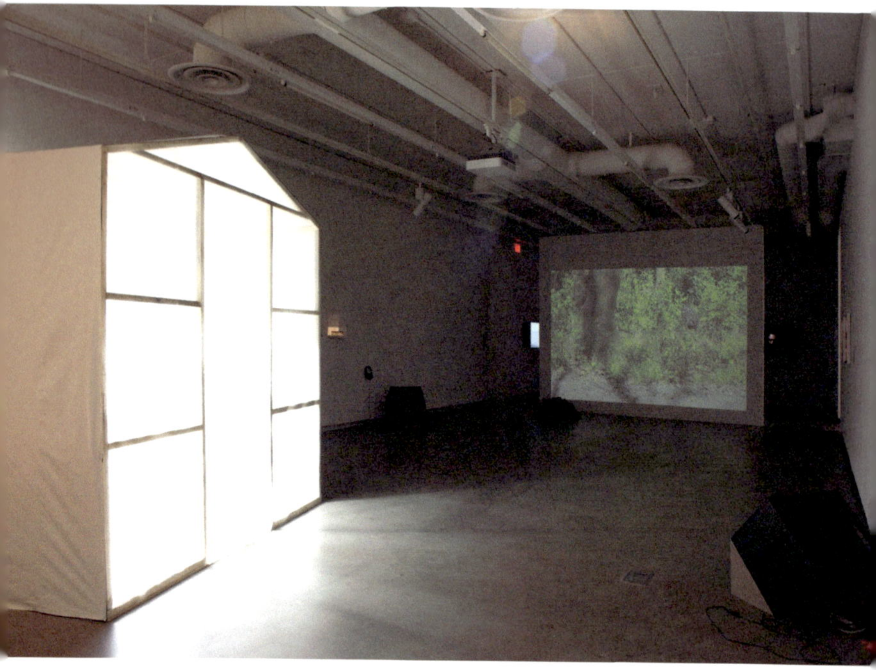

Melentie Pandiolvski & Jennifer Willet, curators, *Toxicity* (Installation view), Plug-in Centre for Contemporary Art, Winnipeg, Canada, 2014. Left: Amanda White, *Frugivore (4th Generation; 2013-2014)*, 2011-ongoing. Indoor greenhouse, lights, seedlings. Right: Alana Bartol, *Forms of Awareness: Ghillie Suit, An Un-Camouflaging*, 2012. HD video, 3 minutes.

endeavours to describe how the world is constituted and experienced through conscious acts. The key idea thus becomes that Phenomenology is able to describe what is given to us in immediate experience without being mediated by preconceptions and theoretical notions, in Husserl's terms a form of transcendental (constitutive) Phenomenology. Husserl named the study of the substance of the mind "phenomenological reduction," a framework that does not assume that something exists, a state that allows positioning the mind for the perception of real, but also absent or imaginary objects. For us, this creates the possibility of linking structures of consciousness through personal experiences in a very wide range of human activities, including those of the arts and technology, and applying them in general ways to the field of Biotechnology, as well as to the intersections of art and biotechnology.

The importance of phenomenological research cannot be overstated when it comes to analyzing the processes characteristic to the intersection of art, biotechnology, and biopolitics. Phenomenology allows us to gain direct experience of the processes informing the artworks, exhibitions, and workshops, before engaging them with critical analysis.

Toxicity

Bioart, as well as ecoart and land art, were all very present in the *BioARTCAMP* project. However, the most surprising for me was the absence of planning for an actual exhibition which would follow the project and interactions. Dialoguing about what was said and left unsaid during the Banff experience led to me and Willet co-curating a big bioart exhibition at Video Pool in Winnipeg (an art venue that I directed). Two years prior to the exhibition I had visited what seemed to be a pristine body of water (Lake Killarney) in Southern Manitoba, by the US border, and was struck by signs warning that cattle should not drink

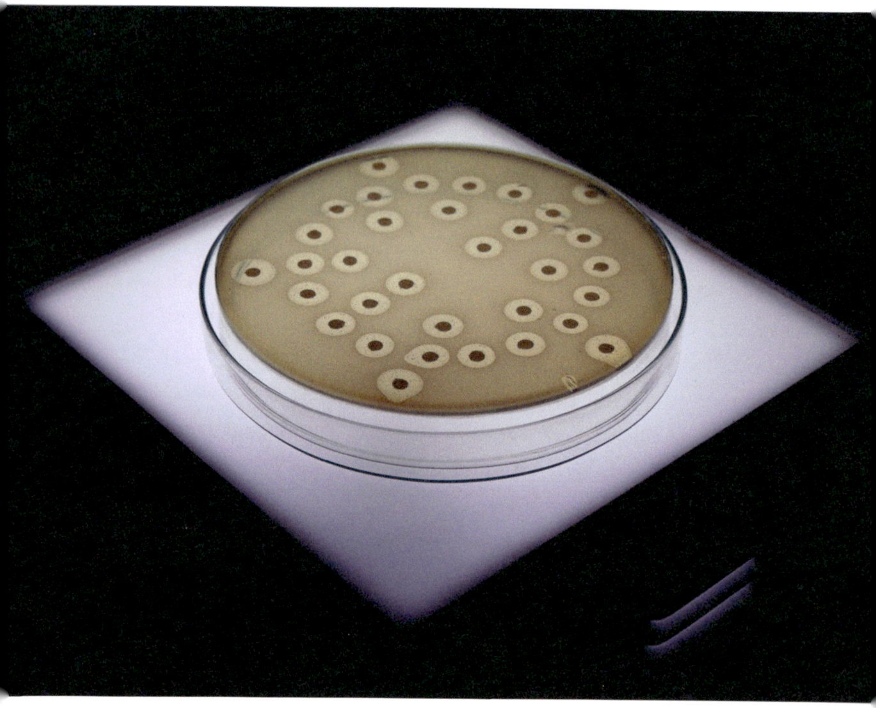

Niki Sperou, *Trust*, 2013. Toxicity testing via zones of inhibition. Glass Petri dish, agar media seeded with E.coli bacteria, antibiotic, paper. Photo credit: Sam Sperou.

the water because it was contaminated. I read this as a sign of destiny and right there I decided that toxicity was going to be the theme of the project. The discussions led to the exhibition *Toxicity*, imagined not only as a showcase of the best bioart practices in the country and internationally, but also as a reconstruction of current environmental situations and socio-political contexts by looking into modes of contemporary cultural and technological production. The project stemmed in many ways from our practical work as curators and artists, but it was naturally also grounded in our academic work.

Toxicity aimed to stimulate cross-cultural dialogue among artists, curators, academics, and the general public at the intersection of art, science, and technology. Another point that became clear from our long individual histories of art-science collaborations was that that artists and scientists addressed the world differently, whether in the long-standing tradition of artists that think beyond the common bounds of cultural standards or the methodological contrasts between artistic freedom and scientific restraint.

Although the idea of *Toxicity* stemmed from my experience of Lake Killarney and *BioARTCAMP*, the exhibition took place at Plug In Institute of Contemporary Art in Winnipeg during the bitter cold winter, from November 2014 to February 2015. In addition to an exhibition, *Toxicity* included a bioart workshop at Video Pool led by Niki Sperou (proclaimed by the participants as Santa Arista Mutata), as well as a public symposium *Toxic Life and Engineered Death* at Winnipeg's Cinematheque, with keynote speakers Joe Davis, Natalie Jeremijenko and Steve Kurtz, and a public screening of *Heaven and Earth Joe Davis* an exceptional documentary by Peter Sasowsky. Other artists involved in the exhibition and symposium included: Trish Adams, Alana Bartol, Joe Davis, Tagny Duff, Aganetha Dyck, Ted Hiebert, Natalie Jeremijenko, David Khang, Steve Kurtz & Critical Art Ensemble, Andrew E. Pelling, Niki Sperou, Reva Stone, Elaine Whittaker, Amanda White, and Jennifer Willet.

Toxicity embedded itself into what could be defined as a standard phenomenological understanding of the co-construction of society and technology in both its theoretical and practical formations. The cultural deciphering of toxic social terrain resonates with current socio-economic global transformations. The implementation of dubious local and international policies, community-based responses, and processes of production, consumption, and disposal are constantly in the public eye. We are continually subjected to processes whose full impact is hard to comprehend but a phenomenological approach allows us to reveal these processes. The Heideggerian *Dinge* (Thing) contains within itself the possibility of gathering together the contents of the universe, toxins included. Relations between Things[5] become crucial, acquiring different features depending on the context, the location, and the reasoning behind the situation. The discourses of biotechnology are evolving, pointing out to us that the latest theoretical and practical developments have a potential to cause a tectonic shift in society and culture, where we now experience the world at the intersection of the engineered and the biological. *Toxicity* appears precisely at this intersection, and its biopolitical modes leave much at stake. Marshall McLuhan noted that the creation of the technological world has created a neural exoskeleton. I maintain that this exoskeleton has become tainted by toxicity in numerous ways, from the bio-environmental to the info-financial.

The extended role of biopolitics today focuses on the crucial question of how biotechnology shapes life and comes to assume a central role in society. Biotechnology, through its complexity, radically reconstructs the relations between politics and nature, allowing for a reassessment of how we look at life today. Under biotechnological pressures, the regulation of life cannot continue under the premises of what has been previously taken for granted. The dualities of power and rights, sovereignty and law, do not escape biopolitics for even a minute. We become witnesses of a process in which the state control of the biological is increasingly being ceded to biotech companies. These companies are sometimes generating their own conflicts (Monsanto and DuPont, for instance, with regard to GMO seed patenting.) In Michel Foucault's words: "For capitalist society it is the biological that is important before everything else; the biological, the somatic, the corporeal. The body is a biopolitical reality; medicine is a biopolitical strategy."[6]

Life, politics and economics intersect at such speeds in the globally-connected society that a novel biopolitical model is emerging which alters this society's operational functions. I refer here to the social and political functions of the Biotech Revolution. The technological and psychosomatic constitute the two poles in this emerging biopolitical discourse. At the centre of this discourse is the notion that life can now be molded as we see fit, enabled by biotechnology. Biopolitics is therefore able to control life by taking it out of the natural domain, reshuffling it at will and subsequently using it in a functional or structural form, thereby freeing life from nature. This intertwining of nature and technology makes the schematic of biopolitics increasingly complex. The question of articulating sovereignty no longer depends on the suppression of life;

our understanding of life and death has altered. Biopower, embedded in biopolitics, now concerns itself with the mere reshuffling of biological units of data. We have moved a step closer to the fulfilment of Foucault's prophecy about the extension of biopower:

> the excess of biopower appears when it becomes technologically and politically possible for humans not only to manage life but also to make it proliferate, to create living matter, to build the monster, and ultimately to build viruses that cannot be controlled and that are universally destructive. This formidable extension of biopower, unlike what I was saying about atomic power, will put it beyond human sovereignty.[7]

Thus, we have effectively banned experiments with plutonium, but we continue to be quite liberal with the experimentation, research, and use of biotechnology. One of the possible outcomes of this arrangement is the creation of a genetically based value system. Another outcome may be the development of appealing forms of neo-eugenics and the creation of new utopian communities. Biology has a long history of being politicized, but we must admit that the biotechnological changes in the past two decades are indeed profound.

Fundamentally, instead of concentrating on producing empirical observations, Phenomenology delves into accounts of experienced space, time, space, body, and relations. Bioart finds good use of all of the above mentioned accounts, but also of one of the basic characteristics of Phenomenology, making a distinction between appearance and essence.

Merleau-Ponty's idea that Phenomenology is the study of essences resounds very well with biotech artworks. Namely, Phenomenology demands the reinterpretation of the world as we interact with it through immediate experience.[8] Phenomenology also insists on a demand for awareness and a will to seize the meaning of the world as that meaning comes into being. Phenomenology always asks after the nature or meaning of something.

The domain of *lived experience* builds on the idea of direct contact with, and experience of, biotech artworks. Specifically, Max van Manen's division of *lived experience* into four distinct categories helps articulate the ways that audiences are able to extract the most meaning from their individual experiences: The *spatiality* of bioart (referring to *lived space*); the *corporeality* of bioart (referring to the *lived body*); the temporality of bioart (referring to *lived time*); and the *relationality* of bioart (referring to the *lived other*).[9]

Lived Space

The concept of spatiality in particular—or lived space (also referred to *felt space*)—represents an integral part of the experience of creating, displaying, and seeing, or witnessing bioart in the gallery space, science lab, workshop area, hospital, or city square.

In this sense of the lived space we can look into Jennifer Willet's *An Incubator in Sheep's Clothing* (2011) or Eduardo Kac's transgenic installation *Genesis* (1999). We can see that they both explore the new fetishes of the biotechnological world: in Willet's case the incubator, a sculpture of a sheep, and live yeast cultures; in Kac's case the gene and the protein, both posing interesting theoretical and metaphysical questions about media, meaning, and representation. The key element in Willet's installation is a functioning incubator placed in the stomach of a life-sized sculpture of a mountain sheep, housing live yeast samples visible through the window. The artist clearly comments on the role of an incubator in

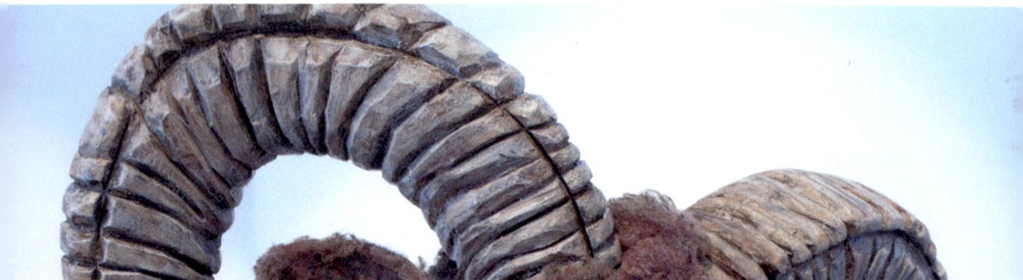

Jennifer Willet, *An Incubator in Sheep's Clothing (detail)*, 2011. Sheep sculpture, incubator, live yeast cultures.

reproducing the conditions of a healthy mammal body for storing and propagating laboratory specimens, as well as on the role of animal surrogates serving as host organisms for interspecies research. In Willet's work we see the ongoing reconfiguring of the relationship between biology and

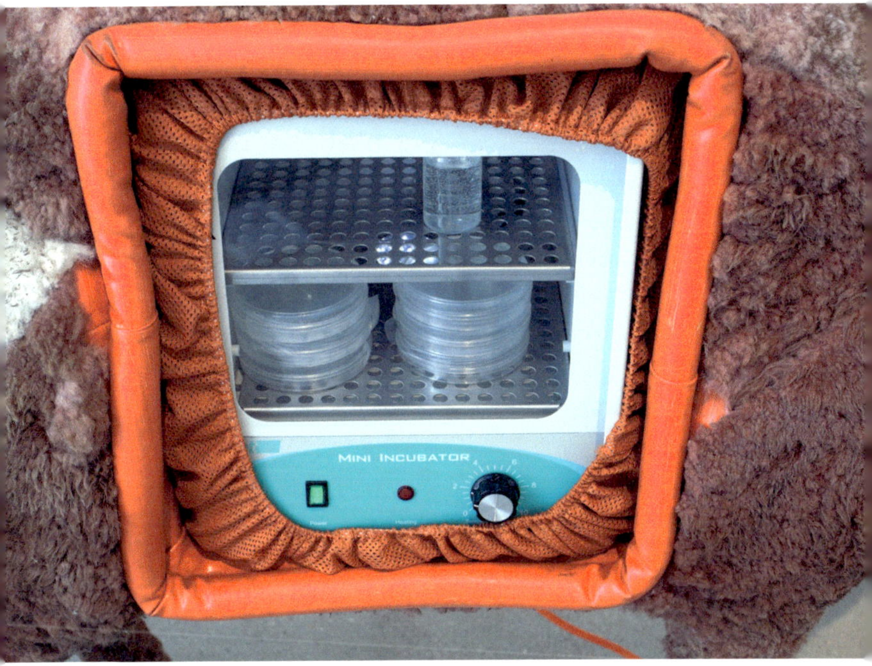

Jennifer Willet, *An Incubator in Sheep's Clothing (detail)*, 2011. Sheep sculpture, incubator, live yeast cultures.

technology; questions of *relationality* are of a high importance as are the questions of lived experience and discourses of the inter-relations, or *lived relations* we maintain with *others*.

The key element in Kac's installation is an "artist's gene," which is a synthetic gene containing a DNA sequencing of the first chapter in the *Old Testament* (the Biblical *Book of Genesis*). The text is translated into DNA bases, and the process is subsequently reversed by translating the mutant gene sequence into to Morse code, and then back to English. Participants in the project (both in the gallery space and on the web) can turn on an ultraviolet light in the gallery, causing real, biological mutations in the bacteria which is presented in a Petri dish in a luxurious glass case. The viewer is thus able to change the biblical sentence in the bacteria itself. This relates to Eugene Thacker's triumvirate of encoding, recoding, and decoding as representing the primary activities of biotechnology today, as well as to the simultaneous notions of the biological stock being property and information, having the traits of materiality and immateriality, and existing as deployments of life that are being shifted from body to body, body to code, and code to body.[10]

Similarly, Joe Davis's work *Malus ecclesia* experiments with a four thousand year-old strain of apple, translated into a seven hundred and fifty million letter book made from the four letters used to encode DNA. During the process Davis inserts Wikipedia entries into the fruit, reproducing the online repository without creating any phenotypic changes in the apple. Similar to Kac's *Genesis* he starts by translating English words into the letters of DNA. Davis's collaborator, the famous biologist George Church, assembles the letters into functional strands of DNA by tricking the bacteria and placing DNA-encoded Wikipedia entries into apple saplings, which are then grafted onto apple stock and allowed to grow into adult trees. This oldest known strain of apple, significantly, is the closest relative of the forbidden fruit that grew in the Garden of Eden.

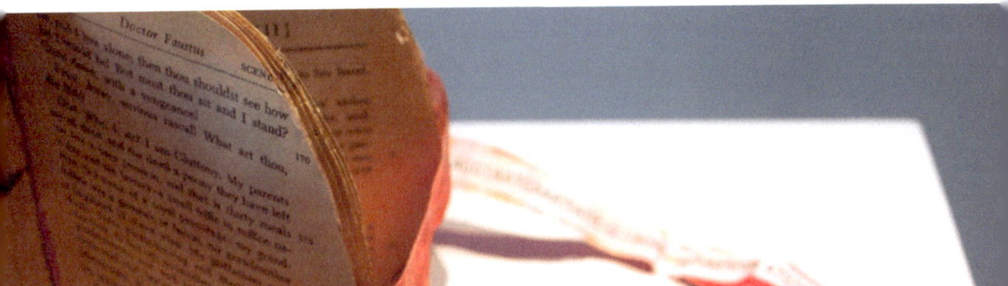

Joe Davis, *Malus ecclesia*, 2013. Mixed media.
Photocredit: William Eakin.

The project thus genetically blends the oldest of apples—metaphors for forbidden knowledge—with the newest archives of freely-accessible electronic information.

Lived Body

The lived body refers to embodied corporeal experience. We experience biotech artworks through our bodies, as we experience the world through our bodies. The experience is always different. For example the viewer of *Kuh,* Michalis Pichler's project, is able to witness a semi-real corpse/artifact in a public space. Namely, *Kuh* consists of tanned skin, claws, and horns of a cow (which was slaughtered for meat consumption) that are sewn over a sculpted corpse carved out of Styrofoam. *Kuh* is then placed in a public space (the work was positioned in different locations in Berlin such as Breitscheidplatz but also in the town of Bruehl). The *lived body* consists of a signifier (skin of dead cow over Styrofoam) and referent (dead cow) makes it a hybrid of an image and a real dead cow. The public touching the artifact has to confront the mortal remains as well as the representation of death.

Joe Davis's *Microvenus* DNA, like all other DNA, is a physical substance. In sufficient quantity, it is visible and tactile. Still, the simple act of seeing something does not automatically make it true. Davis used Sanger sequencing, gel electrophoresis, and autoradiography to verify that *Microvenus* consisted of the intended nucleotides and these techniques produced unique, *Microvenus*-identifying artefacts that have been exhibited internationally.

For Davis, the search for the secrets of life is a search for invisibility within invisibility.

> Details about cellular signalling, membrane transport, vectors, introns, exons and RNAi, promoters and origins of replication are all nested within what there is to know about bacteria. This knowledge includes understandings of transcription of nucleic acids and translation and synthesis of peptides and protein and of the enzymes and biochemical operators that underlie these processes. The study of bacteria is also a study of many useful proteins that assemble into such things as nanoscale molecular motors and fluorescent markers. Bacteria can be assembled into macroscopic devices too, to clean wastewater or produce electrical current. Perhaps most importantly, the study of bacteria is the study of the incredibly compact master plan for life itself that is coded into substances found within each invisible bacterium. The study of bacteria then becomes a desire to acquire the tools and skills needed to resolve this master plan and the wisdom needed to comprehend it. The choice is to learn all of this or to remain forever ignorant. To peer through microscopes is not enough. The facts at hand are much too deeply invisible. The fabric of reality is much finer and far more intricate than the contexts of even the tiniest objects seen in the compass of a microscope objective.[11]

Davis believes that *embodied experience* is not a reasonable limitation, nor a basis to argue that art cannot be created and observed at microcosmic scales. Davis thinks that he has changed the way in which we are asked to behold a work of art.

> *Microvenus* was an image written as language. It is a word and a picture and I created it from scratch. It has since been reiterated in diverse forms. Critics and curators are disoriented because *Microvenus* itself cannot be found perfectly nested in a picture frame or otherwise comfortably disposed in the halls of a typical exhibition. You might ask, "Where are the paints and canvas? Where is the gouache or charcoal? Where is the ink and paper? But, these questions are only asking after the materials I use. How can anything be art if I have not decided to choose from this hallowed inventory of art supplies? I say art is not defined by what substance it is made of any more than it is by the commercial and institutional architectures of galleries and museums. I say this word created for you is more robust and durable than all the pages of all of the history books and the combined collections of world museums. When all of these have finally turned to dust and all of the cities and monuments raised by enormous acts of labour and sacrifice have long since passed from the face of the Earth, this word—and others like it—will be left to shoulder the legacy of human dreams and aspirations.[12]

Lived Time

Lived time refers to the experience of temporal passage, subjective time as opposed to objective time. This notion can be situated within an artwork, particularly in relation to its concept, but also within the audience. With biotech artworks, the questions which are of special interest are those involving the sense of embodied temporality that we face as individuals. It is also useful to look at the concept of stillness as one that has been presented in biotech arts. Eugene Thacker argues that the discipline of bioart requests the creation of stillness, in the physical practices of preservation, as well as a historical stillness of defining and re-presenting the past.[13] Thacker insists that aliveness implies a relationship of dynamic change and relative stillness. In bio-complexity and systems biology they call it a steady-state. The concept of lived time via the notions of *stillness* and *aliveness* also includes possible notions of speeding up, or slowing down of our perceptions of time. We can even relate this concept to the time wasted by audiences that do not want to stay in an exhibition for a long time, but are compelled to do so in the process of trying to understand the work

Relationality

Relationality involves numerous sorts of interrelationships. Max van Manen, in *Researching Lived Experience: Human Science for an Action* speaks of the lived relationship we maintain with others in the interpersonal space we share with them. Forms of relationality include seeking to establish a shared presence in a single space involving the *lived other*. Discourses of inter-relation involve any others who were present when communicative transmissions took place, such as other artists, the general public, mechanical and electrical engineers, biologists, astronomers, professional dancers, architects, linguists, and philosophers.

One such experience is one I termed *magnetic baptism,* a performance I organized during the *Toxicity* project, in a curated pseudo-scientific ritual, after experiencing Ted

Hiebert's installation, by placing strong magnets on both sides of Hiebert's head. In his project *Between Magnets* for the exhibition, Hiebert patiently allowed these strong magnetic fields to affect his body and consciousness. We did not know what the consequences might be, as we simply

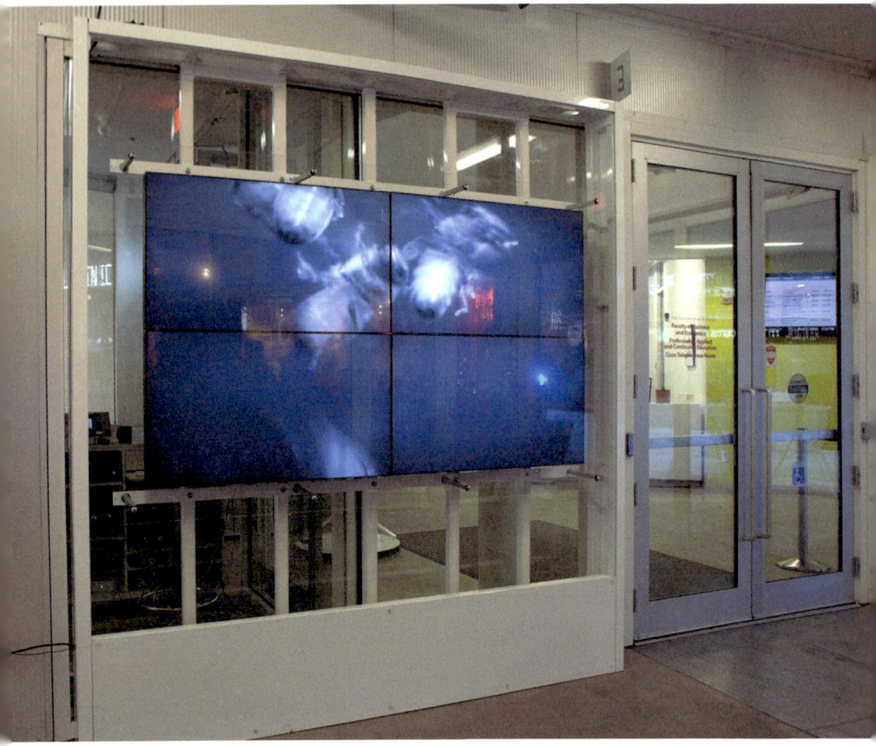

Trish Adams, *Disordered Swarming*, 2013. Video installation with QR code access. Plug In Institute of Contemporary Art, Winnipeg, Canada. Photo credit: William Eakin

do not know enough about magnetism in general. After all, Albert Einstein spent decades attempting to develop the *unified field theory* which would relate electromagnetism and gravity. But he was only partially successful. However, a posthuman artistic proposition is bound to consider advances in understanding electro-magnetism, in the same way it embraces biological sciences.

Similarly, Trish Adams's artwork *machina carnis*—the last work in a series of works called the *vital force* series—inquires about corporeality and potential changes to the body as we know it today. In a quasi-scientific way, her artworks reference and parody the development of galvanics at the beginning of the nineteenth century. This experimental series also incorporates the artist's curiosity about the ways in which the discovery of electricity led to a spate of new mechanical devices for measuring the physical responses of a human body. She is intrigued by the way that scientists of that era appeared to feel that this new technology (galvanics) would at last provide the key tools to quantify and explore the body and thus finally measure the essence of "humanness." Of course we are able to see parallels here with our own latest technology for measuring the body: the sequencing of the human genome.

The Phenomenology of (Non)Habitual Spaces

The disclosure of the world through technology is also a disguise of the relations technology is used to build, relations that can, however, be unconcealed. In *The Question Concerning Technology*, Heidegger noted that the essential unfolding of technology harbors within itself what is least expected: the possible rise of a saving power.[14] Where does this saving power of technology reside? Will salvation be found in art and activism, that is, strategies of resistance? Perhaps the Biotech Age will allow us to witness what Heidegger called *the second beginning of thinking*, the meeting of the world in historical time-space. And, perhaps this space can be built only by art. Artistic and cultural research into biotechnology has questioned established philosophical systems, ethical beliefs, and cultural practices by proposing new ways of looking at life and society, as artists, critics, and theorists navigate the maze of the Global Biopolitical Apparatus. How art and technology interrelate and how biotechnology infiltrates everyday life—and how these interrelations change the cultural, sociopolitical and ecological landscape—are becoming increasingly important research areas. Artistic responses to these questions have been vast, including: the examination of biopolitical conflicts in real and virtual worlds; pollution; corporeality and somatic biopolitics; energy control, fuel material and alternative energy sources; electro-magnetism; the inheritance and programmability of life; the causes and consequences of environmental changes; environmental sustainability; micro and macro-ecologies; life, empathy and questions of ownership; GM products; death and appearance; and the ethical implications of working with biological media in an art context. In many ways these artistic responses are also practical strategies of resistance, ones that we need in order to address existing structures of knowledge and to achieve broader ethical and philosophical considerations of biotechnology. We must look into what Heidegger would have called the Biotechnological Gestell (Enframing) of everyday life and address the changes caused by toxicity in the cultural, socio-political and ecological landscape.

Notes

1 BIOTEKNICA was a not-for-profit artist collective founded by Jason Knight and Jennifer Willet. It was active from 2000-2007. Its purpose was to critically investigate the ethics, aesthetics, and technological potential of new art forms that lie at the intersection of the arts and the biological sciences.

2 The *Art of the Biotech Era* group exhibition was presented at the Experimental Art Foundation in Adelaide, Australia, in 2004. Eduardo Kac also held a workshop in Adelaide in 2005. The project was continued with *Biotech Art Revisited* in Adelaide in 2009.

3 The installation *Genesis* focuses on bacteria that contain a DNA sequence of the *Book of Genesis*. The key element of *Genesis* is what Kac calls an "artist's gene," a synthetic gene created by translating biblical material into Morse Code, and converting the Morse Code into DNA base pairs. The "Genesis" gene is incorporated into glowing bacteria and projected as live video in the gallery and streams over the Internet, where the public is encouraged to intervene and monitor the evolution of the work.

4 *Alba—The GFP Transgenic Bunny* was created in 2000, at the INRA Institute in France. While an embryo, French scientists inserted the jellyfish gene that produces Green Fluorescent Protein (GFP) to make Alba glow.

5 We see here the important shift that Heidegger makes from Object (*Gegenstand*) to thing (*Das Dinge*). This is not merely a word play but a paradigm shift that takes place where the Objects of Science, Technology, and of Art, for that matter, are viewed as things. The integrity of Aristotelian substances is broken down in this way, as *Dinge* has encoded into it the possibility of gathering the contents of the Universe. This position results in a profound change, as Aristotle's belief that relations to other objects are a mere accident and leave the essences of objects unchanged can no longer be applied. The relations between Things become

crucial. This position essentially means that Things have different features according to where they are situated and the context in which they are placed.

6 Michel Foucault, *Society Must Be Defended: Lectures at the Collège de France, 1975-1976*, David Macey, trans. (London: Verso, 2003).

7 Foucault, *Society Must Be Defended.*

8 Merleau-Ponty in the *Preface* to the *Phenomenology of Perception* points out that the work of Phenomenology is as painstaking as the work of artists such as Balzac, Proust, Valery, or Cezanne.

9 Max van Manen, *Researching Lived Experience: Human Science for an Action-Sensitive Pedagogy* (London: Routledge, 1998).

10 Eugene Thacker, *The Global Genome: Biotechnology, Politics, and Culture* (Massachusetts: MIT Press/Leonardo Books, 2006), 305-320.

11 Melentie Pandilovski, Interview with Joe Davis, Skopje, Macedonia, 2010.

12 Pandilovski, Interview with Joe Davis.

13 Thacker, *The Global Genome.*

14 Martin Heidegger, *The Question Concerning Technology and Other Essays*, William Lovitt, trans. (New York: Harper Torchbooks, 1977), 337.

11 Notes From the Deep Earth Treatment Centre

Amanda White & Alana Bartol[1]

We are all compost, not posthuman.

— Donna Haraway[2]

Soil may seem at first to be an unglamorous topic, yet its contradictions make it most interesting; its familiar, ubiquitous nature allows it to be easily dismissed as "dirt" and yet it has simultaneously been described as "the most complicated biomaterial on the planet."[3] Most simply put, soil is a combination of organic, clay, and mineral materials constituting the top-most layer of earth where plants grow. Dig a little deeper (so to speak) and it is a much more complicated, vastly misunderstood, and largely unknown collective of elements that form the basis of all life on earth. While it is fascinating to note that "in one single handful of fertile garden soil there will be more organisms than the total number of human beings that have ever lived,"[4] it is also extremely difficult to imagine or visualize such numbers; they are so large that they become incomprehensible. In the introduction to a special issue of the journal *Science* on the topic of soils, Andrew Sugden, Richard Stone, and Caroline Ash describe these mysterious qualities aptly: "In many ways, the ground beneath our feet is an alien and distant planet, the processes occurring in the top few centimetres of earth's surface form the basis of all life on dry land, but the opacity of soil has severely limited our understanding of how it functions."[5] The opacity described by the authors could be thought of as both the biological complexity of soils and the ways in which many of its processes are microscopic or invisible to the naked

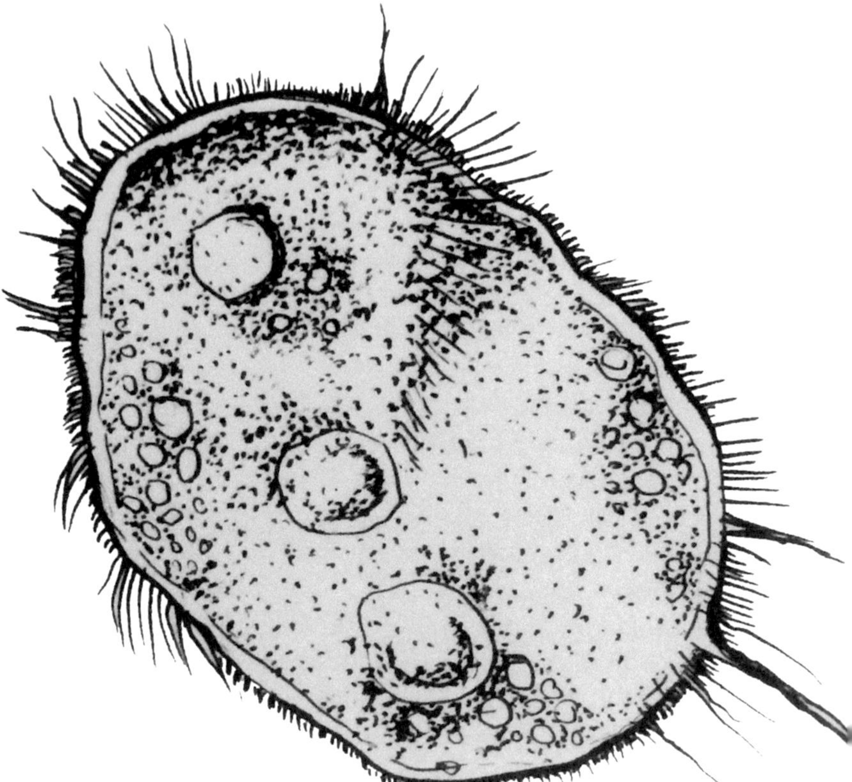

Amanda White, *Microorganism*, 2015.
Ink and watercolor on paper.

(human) eye, which together with its scale make soil nearly impossible to consider from a human perspective.

The unknowable, alien nature of soils depicted here is of course attractive to our wilder imaginations, however these facts emerged later for us. Our initial interest in soils began more modestly, after a scientific study linking mental health with soil microorganisms was brought to our attention. This study—which yielded positive results when performed on

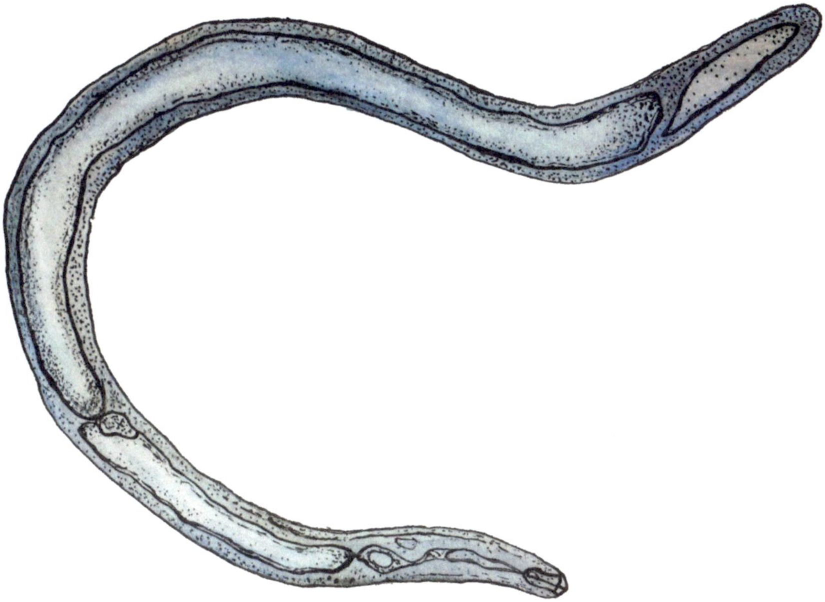

Alana Bartol, *Nematode*, 2015.
Ink and watercolor on paper.

mice—proposed that physical interactions with a particular soil microorganism (*Mycobacterium vaccae*) would trigger a release of serotonin in the brain, causing elevated moods and decreased anxiety[6] much like antidepressant pharmaceutical drugs.[7] The researchers—Christopher Lowry, Jacob Hollis, Annick De Vries, and others—hypothesized that physical contact with soil may hold similar antidepressant properties for humans. We found this interesting for several reasons: first, it was our introduction to the diversity and abundance of life that exists in soils and how much of it remains virtually unknown; and second, this study presents an example of the scientific process proving something that is colloquially known: that gardening is good for you. This is an excellent example of how ways of knowing can intersect or how different forms of knowledge inform and legitimize one another, instances that we seek out in our artistic practices as we work and collaborate.

Perhaps not surprisingly, in attempting to explore the topic of soil further, we have been met with a particular roadblock: much of what is known about soils (and their resident communities of microorganisms) is based on the economic value of soil as an agricultural commodity. Our findings suggest that most questions currently asked about soils are of the "how can we use it?" or "how might it benefit us?" variety. The study by Lowry et al. deviates from the narrative of soil as agricultural resource yet still remains rooted in the understanding of use-value for humans.

Amanda White, *Microorganisms*, 2015.
Ink and watercolor on paper.

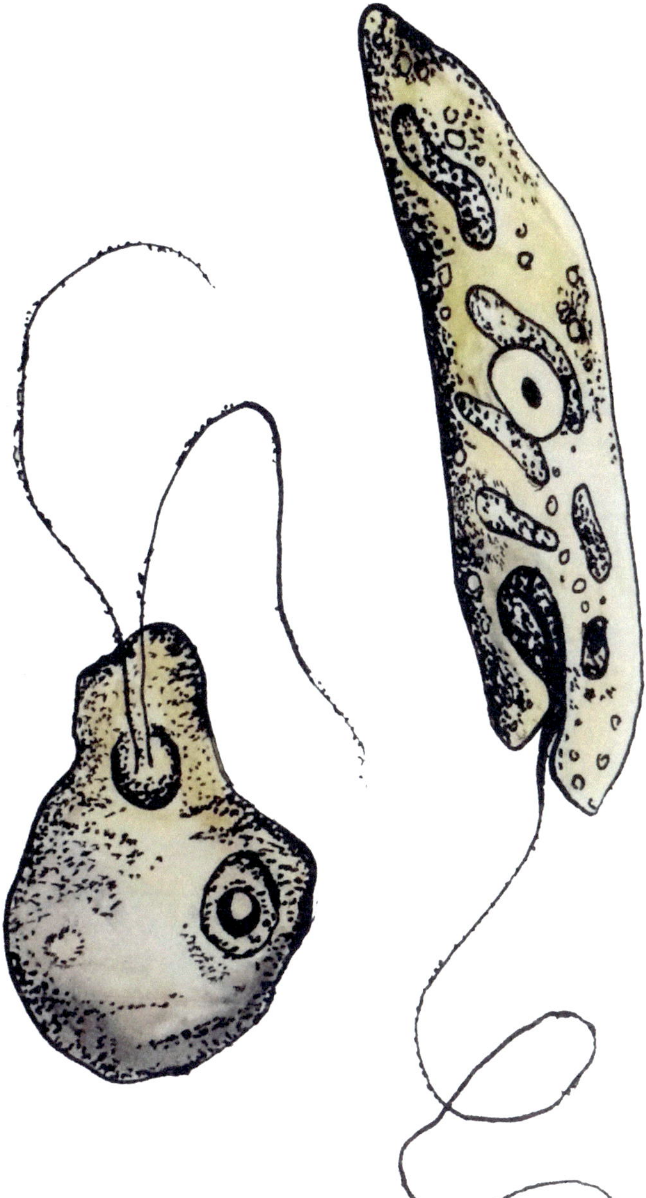

In 2014, we co-founded the Deep Earth Treatment Centre (DETC) as an artistic platform from which to stage experiments, interactions, and interventions exploring the connection between humans and soil. In the beginning, our own work at the Deep Earth Treatment Centre was also focused on the therapeutic potential of such interactions for humans alone. Our first DETC project took the form of a guided visualization using transformative imagery, working with a group of students in a bioart course at the University of Windsor. Each participant was given a pillow filled with organic soil to lay or sit on. In our journey we became earthworms, moving through the soil and creating airy tunnels. As the smell of earth filled the room we encountered the soil in our mind and body, yet the focus was still on our human experience.

We have since taken this bias up as a challenge to be engaged, and are determined to push the limits of a normative anthropocentric understanding of soils. As Donna Haraway suggests in her recent book *Staying With the Trouble: Making kin in the Chthulucene*, "The task is to make kin in lines of inventive connection as a practice of learning to live and die well with each other in a thick present."[8] In the spirit of living together and working across species in difficult environmental times, we attempt to stay with this particular trouble by asking a question: If soil can be shown to contain anti-depressant properties for humans, should we not also ask what makes soil happy? What does soil want? Does human physical engagement with soil have a positive effect on soil health? By attempting to open this direction of the dialogue, we are expanding our imagination of soils beyond that of a resource for human exploitation—be it economic or therapeutic—towards an understanding of soil as a complex, multi-species community with which our relationship may take many forms and have many mutual benefits.

While we seek new entry points to the understanding of soils, the fact that human-soil symbiotic relationships appear most obviously in agriculture does not interfere with the project of approaching it anew with an ethics of care. Indeed, this may instead present an argument for the necessity of such a shift, as the anthropogenic depletion of soils due to industrialized farming practices and urbanization has been identified as an urgent global issue. In fact, one could argue that the continued disregard of the importance of soils is dangerous, posing a threat to both biodiversity and global food security. Activist and ecofeminist scholar Vandana Shiva writes about the current crisis in her book *Soil not Oil*, proposing that we re-imagine our relationship with soils as an alternative to the attitude of "deadly consumerism" and instead as an opportunity to "become co-producers and co-creators with nature,"[9] articulating that "soil teaches us to be earth citizens."[10] Shiva's conceptualization respects the soil as a bearer of knowledge, elevating its position to that of a teacher rather than an object or commodity. Biologist Lynne Margulis (well known for her work with James Lovelock on the Gaia hypothesis) similarly denounces human exceptionalism by envisioning our relationship to soils (and to all non-human cellular organisms) differently. Margulis writes in her book *Symbiotic Planet* that "while humans have indeed evolved, they have not done so independently—solely from apes or even from other mammals—but from a long line of progenitors and ultimately from the first bacteria."[11] Margulis's proposal of distant and microscopic kin suggests a consideration of time on a planetary scale and a familial kind of human-soil relationship, unacknowledged in our current configuration of species hierarchy and categorization in which the tiny, invisible life in the soil is so often rendered inconsequential.

At the Deep Earth Treatment Centre we are similarly critical of human-centered world views and seek alternatives to the often one-sided flow of understanding as an approach to the exploration of, not just what we don't know, but also how we might be of benefit to soils. At the DETC, the subjects of our work are both humans and soils, and the "treatments" are a kind of mediation tending to this relationship through physical, affective, and sensory explorations.

In 2016, we embarked on a 5 year research-creation project in collaboration with Jennifer Willet, entitled *BioArt: Collaborating With Life*. The work we are developing is hosted at Willet's bioart laboratory at the University of

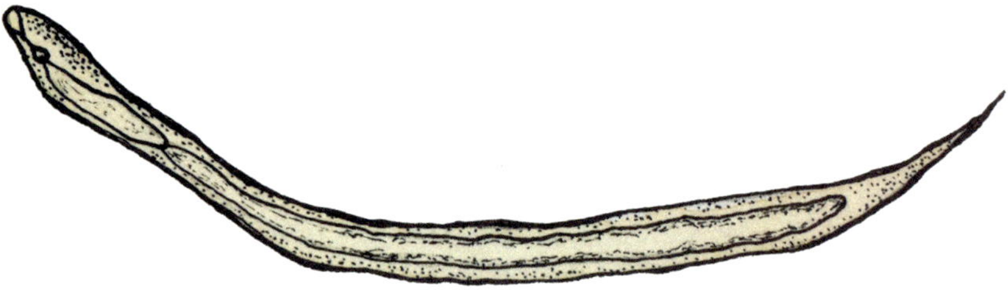

Alana Bartol, *Nematode*, 2015.
Ink and watercolor on paper.

Windsor and connects our work as three artists interested in ecology, participatory art practices and working across species. Willet's body of work to date has focused on the kinds of unique ecologies and interspecies relations that are formed and cultivated in laboratory environments; her bioart laboratory—INCUBATOR Hybrid Laboratory at the Intersection of Art, Science and Ecology—re-imagines a space of scientific inquiry as one for artistic experimentation and public accessibility. From the scientist herself to microscopic bacteria in a petri-dish, Willet's work implicates every organism in the laboratory as part of its ecology or community, creating spaces of democracy from those traditionally embedded with hierarchies of gender, species, and knowledge. In this lab, science is not performed in closed sterile rooms by men in white coats, but instead conceptualized as messy, playful, multi-species encounters. Our collective project aims to focus on the possibilities, issues, and problems that arise when working across species through artistic production and research. Working in Willet's INCUBATOR Lab lends itself to the application of more science-based methodologies to our work, using tools such as microscopy imaging, sample collection, and testing to approach questions about soils in new ways.

The irony of our title, *Collaborating with Life*, does not escape us as it is a contested idea that humans might "collaborate" with non-human others, particularly those with whom we cannot easily communicate. This is a core concern for us; indeed we may prove the very impossibility of such collaborations through this work. However, opening dialogues around these ideas is part of the goal of the project. As such, our obligations to the more than human constituents whom we work and considerations of care for them have inevitably come to the fore as this work develops. Maria Puig de la Bellacasa writes thoughtfully on considering care beyond the human in her book *Matters of Care: Speculative Ethics in More Than Human Worlds*,[12] referring to Joan Tronto's definition of care as "everything that we do to maintain, continue and repair 'our world' ... which we seek to interweave in a complex, life sustaining web"[13] and further suggests that there is a necessity in exploring the "significance of care for thinking and living in more than human worlds."[14] Puig writes that "caring is an effective state, an ethical obligation, and finally a practical labor,"[15] referring to a feminist ethics of care and to others such as Donna Haraway who also apply such considerations to non-humans. In an analysis of the concept of care as it is taken up by Haraway and Puig, Thom van Dooren remarks that for both theorists, care becomes a central feature of living with and becoming kin in multi-species worlds, that their work suggests an embodied practice, summarizing that care "requires that we DO something wherever possible, to take care of another."[16] He then asks a poignant question: "What does it mean to care for others at the edge of extinction? What forms might careful scholarship take at this time?"[17]

Surely, when engaging with other-than-human subjects in critical environmental times there may be obligations in addition to the common ethical considerations that always arise when doing community-based work or research. In models of participatory action research for example, issues of power, ally-ship, and hierarchy arise and can be identified and addressed through tried and tested methods. Attempts are often made to acknowledge and prevent or repair such imbalances through collaboration, where researchers work in partnership with their subjects.[18] In a methodological experiment from 2013 titled "In conversation with ...: co-designing with more-than-human communities," a group of scholars in the United Kingdom sought to apply the principles and methodologies of participatory research to other-than-human research subjects in a series of interactions as a "speculative"[19] examination of the transferability of these models. Describing the results of this methodological experiment, lead researcher Michelle Bastian concluded that when working across species, unique complications arose, such as "issues of ethical relationality, the problem of representation, of exchange across different perceptual worlds and anthropocentrism constituting some of the area's most pressing issues."[20]

The problem of relating is not confined to academic projects either, as similar issues are encountered by artists working across species using participatory methodologies. In socially engaged artwork for example, identifying non-humans as participants or even collaborators in a work is often used as a strategy for inclusion,[21] yet this framework may raise the same questions of ethics and representation as those identified by Bastian in her related research.

Additionally, speed and time present hurdles, as other-than-human speeds are part of the different perceptual worlds encountered in such projects. Geologic time, plant-time, and in this particular case "soil-time"[22] should be considered if we intend to adopt an ethics of care in this work. While this is both necessary in order to approach understanding, and a responsibility that we have to our soil participants, putting these concepts into action presents ongoing challenges. There is hope in the current turn towards slow scholarship, which asks researchers to adapt a "care-full"[23] approach to their work. Slow scholarship is suggested as a form of resistance to neoliberal practices in the university, where demands of fast turnover for research and production exist[24] (also features of the art world). The goal of this movement is to have a more thoughtful approach to time, "to slow down and spend more time with subjects; advocating for good scholarship requires time: time to think, write, read, research, analyze, edit, and collaborate."[25] This move towards the adaptation of a feminist ethics of care in community-based research is in line with developing a similar model of care when working across species in research and art practice, particularly in our contemporary moment of ecological crisis.

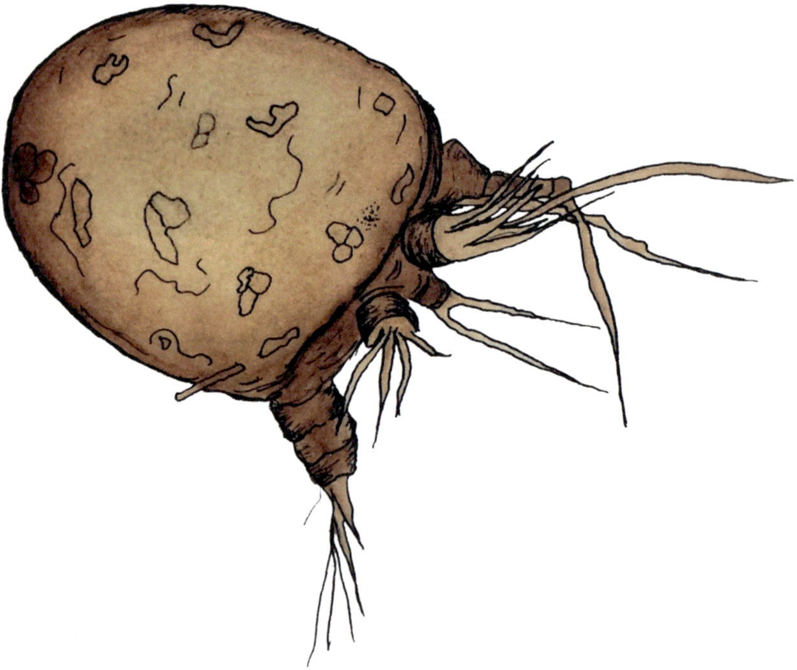

In March 2017, the DETC and Willet gave the first public presentation of our collaborative work; a three-day event called *Life in the Soil.* We felt that a focus on soils—the unassuming, overlooked, and yet vitally important element—would be an appropriate starting point for our research on the topic of *Collaborating with Life*, to begin with our bacterial ancestors, micro-organisms and life in the soil.

The event featured workshops, performances, and participatory discussions with invited artists, activists, farmers, and scholars invested in scientific, indigenous, embodied, and practical knowledge about the vital nature of soils. Overall, the cross-disciplinary dialogues that formed *Life in the Soil*, with its many different voices, was a means to encourage various ways of knowing, towards deepening our relationship with and understanding of the complexity of living soil and its importance to all life on earth. A raised bed of soil formed the center around which these many discussions, performances, and happenings took place over the three-day event. We conceived of the soil bed as a participant for whom we needed to make space, who we must interact with, in order to confront soil physically rather than solely conceptually. Human participants were invited to create actions in response to the soil over the three days: artist and Reiki practitioner Kacey Auffret performed a Reiki treatment for the bed; Alana Bartol slept overnight on it; I buried myself up to my belly in the bed and read a children's story to my then unborn baby, to name

Alana Bartol, *Soil Mite*, 2015.
Ink and watercolor on paper.

a few examples. While the presence of the soil bed aided in heightening our own awareness of life underground and enabled more meaningful physical interactions for us, its presence also re-confirmed the limitations of such an approach; for example, the issues of soil-time and agency were not resolved. It is difficult to discern whether the actions performed for the bed had any effect, though we have collected and are currently testing samples from the bed to address this, to see whether there are any changes in the general properties of the soil before and after the event (how we might interpret these results is yet to be determined). For now, the soil retains its mysterious identity as both the substrate of our existence and a distant, alien planet.

The driving force underlying our work as the DETC is the simplest, and yet perhaps most difficult question to answer: how might we get to know the life in the soil? Some of the preliminary approaches we have taken are shared in a bit more detail in the following "Selected Records." Often collaborative and speculative, what each of these projects points towards is our ongoing commitment to thoughtful engagement with and around the earth.

The Deep Earth Treatment Centre: Selected Records

The Dream Experiment—January 2015

During the "Food Water Life" residency with Lucy + Jorge Orta at The Banff Centre for the Arts and Creativity, the Deep Earth Treatment Centre collected soil from three significant sites in Banff: Tunnel Mountain, Sulphur

Mountain and The Banff Cemetery. Over the course of a week, participants in this experiment were asked to place a color coded test tube filled with soil from one of these (undisclosed) locations under their pillow and record their dreams. At the end of a week, a gathering was held during

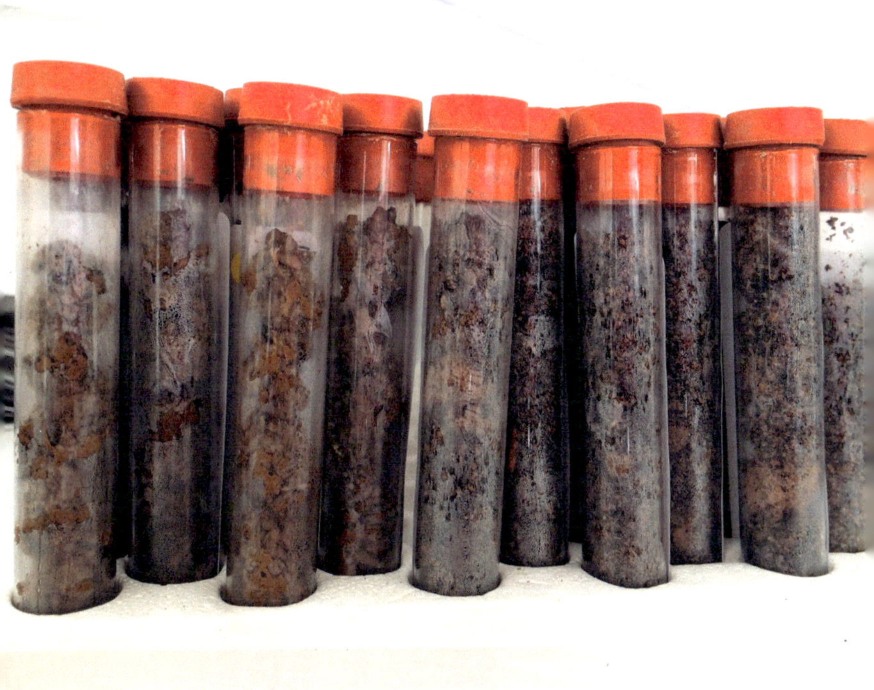

Deep Earth Treatment Centre, *Soil samples collected in Bannf, Alberta*, 2015. Photo credit: Alana Bartol.

which the dreams were shared with the group and the soil locations were revealed. The soil was mixed into a tincture and returned to the sites where it was collected, embedded now with many dreams.

Life in the Soil—March 2017

Life in the Soil was a three-day event hosted by the INCUBATOR Lab at the School of Creative Arts, University of Windsor, featuring interactive workshops,

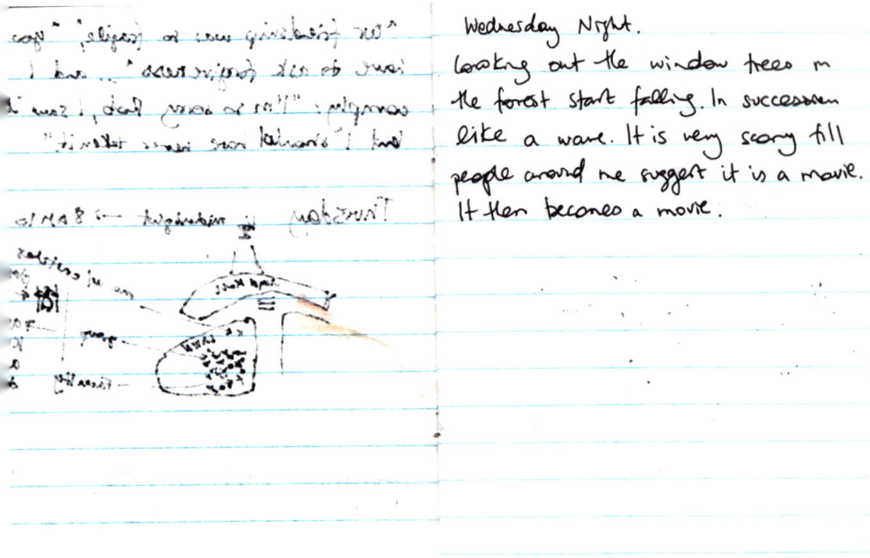

A participant's dream journal from the *Banff Dream Experiment*, 2015.

excursions, and participatory discussions led by artists, activists, farmers, and scholars invested in the scientific, indigenous, embodied, and practical knowledge about the vital nature of soils.[26]

In one of many events during *Life in the Soil*, Dr. Maria Cioppa (University of Windsor, Environmental Sciences) discussed the interplay between Essex County's soil, climate,

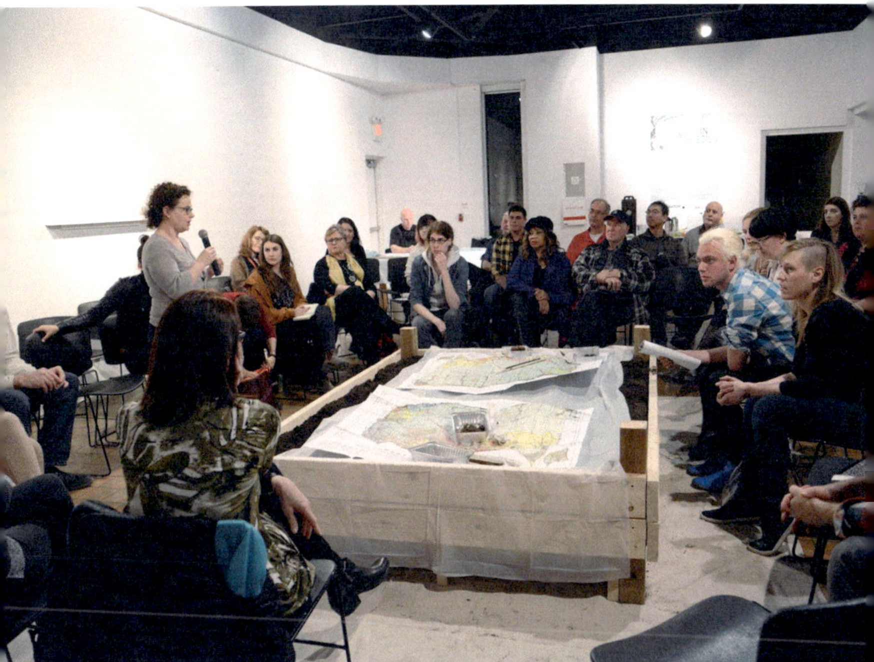

Incubator Lab, *Life in the Soil*, 2017.
Event documentation, photo credit: Shallen Chen.

and wine, while a tasting of local wines was facilitated by grower Scott Wilkins and Adam Graham from Coopers Hawk Vineyard.

Earth Moves was a movement workshop developed by choreographer Ahn Nguyen (Windsor) and HNM Dance Company. Short performance pieces were presented by

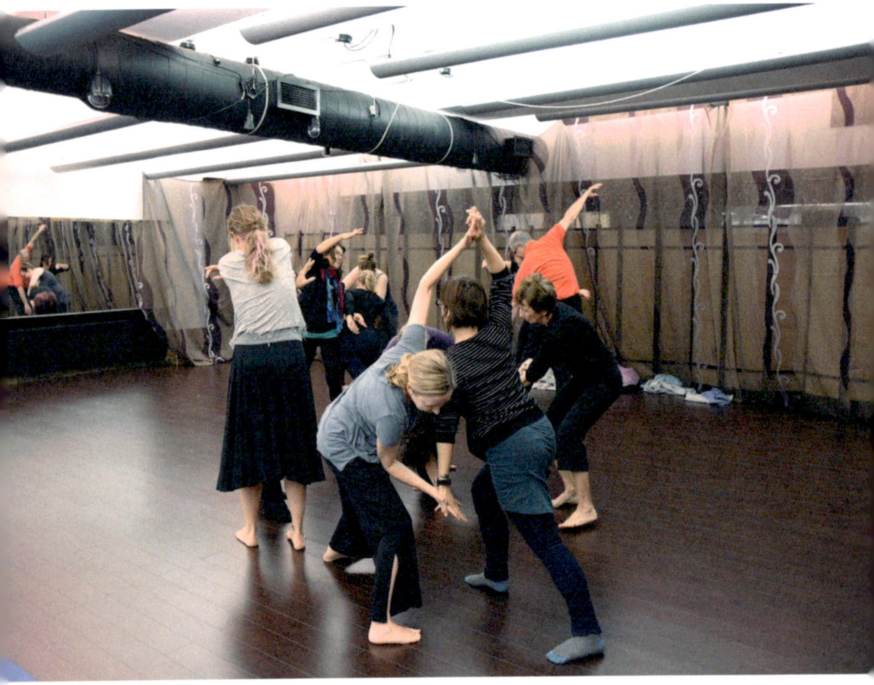

Ahn Nguyen, *Earth Moves*, 2017.
Event documentation, photo credit: Shallen Chen.

the company and participants were invited to collectively explore the vast world below the earth's surface through guided movement exercises.

Soil Bed was a participatory installation created by the Deep Earth Treatment Centre at the University of Windsor's Lebel Gallery. The work was comprised of locally

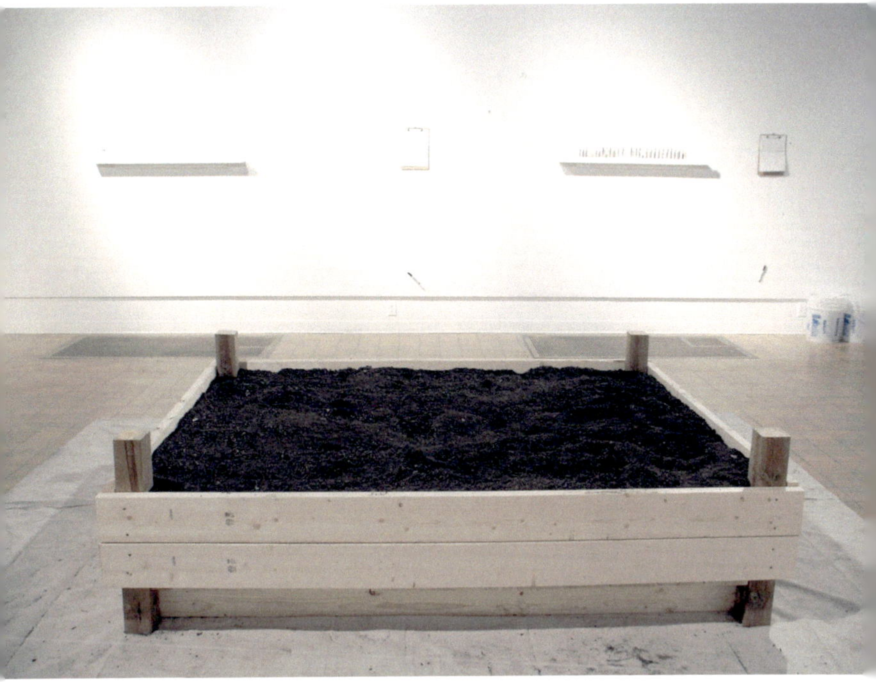

Deep Earth Treatment Centre, *Soil Bed*, 2017.
Installation view, photo credit: Shallen Chen.

purchased commercial topsoil placed in a wooden frame the size of a standard queen bed (60" x 80"). Before the event, the DETC invited participants to collect soil samples from "contested" sites anywhere in the country. This was open to interpretation. Over the course of the event, participants shared where and why they collected their samples.

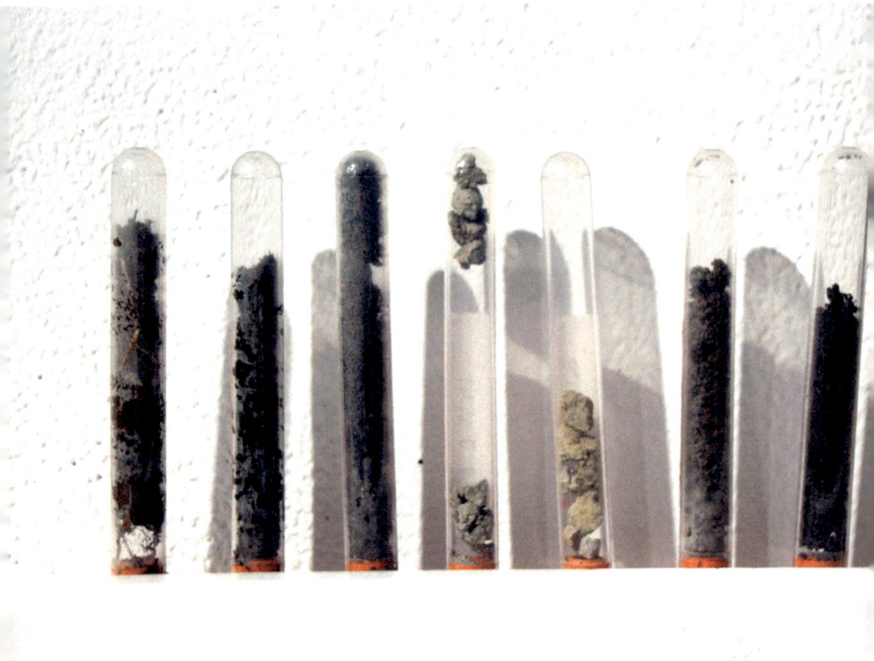

Deep Earth Treatment Centre, *Life in the Soil: Soil samples collected by participants from contested sites*, 2017. Photo credit: Alana Bartol

On the first day of *Life in the Soil*, each participant's soil sample was mixed into *Soil Bed.* Audiences and artists were invited to engage with and respond to the bed of soil through experimental gestures, actions, or exchanges as symbolic or actual steps towards remediation. Responses included sleeping on soil, giving a Reiki treatment to the soil, burying

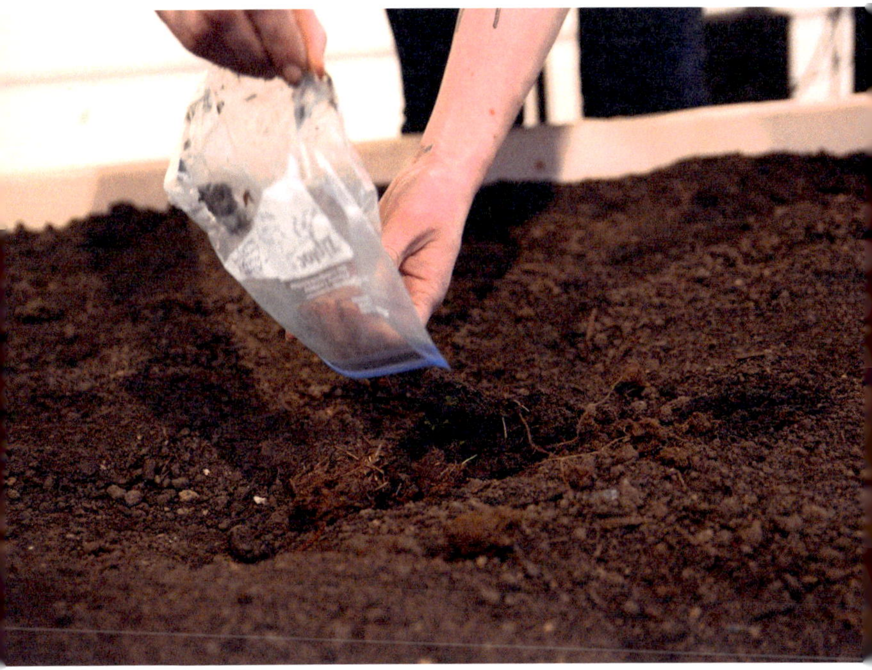

Deep Earth Treatment Centre, *Life in the Soil:* Soil samples collected by participants from contested sites are mixed into *Soil Bed*, 2017. Photo credit: Shallen Chen.

objects, an expecting mother reading to her unborn child while immersed in the soil (womb reading), and other forms of physical contact with the soil, while the bed also formed a centerpiece for the three-day event. Samples taken from the soil bed—before and after engagement—are currently being tested in a laboratory for measurable differences.

On March 8, 2017, Alana Bartol spent the night sleeping on Soil Bed, which had 15+ soil samples from contested sites, including sites with real or perceived contamination. Building on DETC research exploring dreaming and cross-species communication, the performance was a gesture towards remediation and empathy, imagining possibilities of exchange with microscopic life in the soil.

Notes

1 The Deep Earth Treatment Centre is a collaborative work by Amanda White and Alana Bartol, "Notes from the Deep Earth Treatment Centre" was written by Amanda White with "Selected Records" by Alana Bartol.

2 Donna Haraway, "Anthropocene, capitalocene, plantationocene, chthulucene: Making kin," *Environmental Humanities* 6.1 (2015),161.

3 I. M. Young and J. W. Crawford, "Interactions and Self-Organization in the Soil-Microbe Complex," *Science* 304.5677 (2004), 1634.

4 Ibid.

Alana Bartol, *Sleeping on Soil*, 2017.
Performance/action, photo credit: Amanda White.

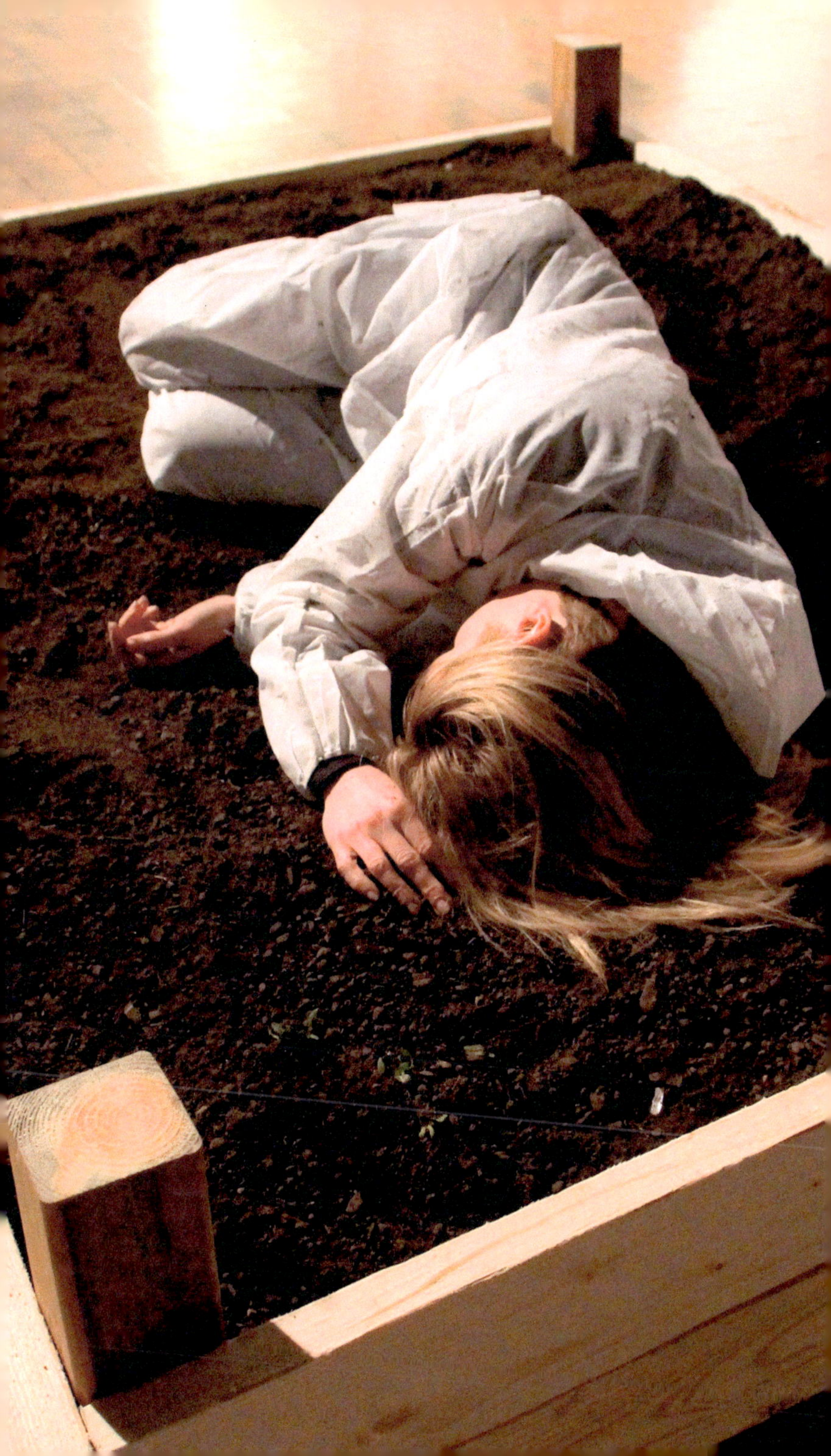

5 Andrew Sugden, Richard Stone and Caroline Ash, "Ecology in the underworld," *Science* 304.5677 (2004), 1613.
6 Christopher A. Lowry, Jacob H. Hollis, Annick De Vries, Baohan Pan, Laura Rosa Brunet, Jon RF Hunt, Julian FR Paton et al., "Identification of an immune-responsive mesolimbocortical serotonergic system: potential role in regulation of emotional behavior," *Neuroscience* 146. 2 (2007), 756-772.
7 Josie Glausiusz, "Is Dirt the New Prozac?" *Discover Magazine*, June 14, 2007. http://discovermagazine.com/2007/jul/raw-data-is-dirt-the-new-prozac.
8 Donna Haraway, *Staying with the trouble: Making kin in the Chthulucene* (Duke University Press, 2016), 1.
9 Vandana Shiva, *Soil not oil: environmental justice in a time of climate crisis* (Brooklyn: South End Press, 2008), 7.
10 Ibid.
11 Lynn Margulis, *Symbiotic planet: a new look at evolution* (New York: Basic Books, 2008), 4.
12 María Puig de la Bellacasa, *Matters of Care: Speculative Ethics in More Than Human Worlds* (Minneapolis: University of Minnesota Press, 2016).
13 Ibid., 69.
14 Ibid., 1.
15 María Puig de la Bellacasa, "'Nothing comes without its world': thinking with care." *The Sociological Review* 60. 2 (2012), 197.
16 Thom van Dooren, "'Care' in the Living Lexicon for the Environmental Humanities," *Environmental Humanities* 5 (2014), 292.
17 Ibid., 291.
18 Michelle Bastian, "Towards a more-than-human participatory research," in M. Bastian, Owain Jones, Niamh Moore and Emma Roe, eds., *Participatory Research in More-than-human Worlds* (London: Routledge, 2017), 19-37.
19 Ibid., 21. The authors described the work as experimental, a "what if" or speculative scenario in which they are

asking what would it mean to include other than humans in participatory research rather than advocating for it necessarily.

20 Ibid., 20.

21 Anthony Marcellini and Matthew David Rana, "Notes Toward a Non-Anthropocentric Social Practice | Art Practical," *Art Practical* 3.11 (March 14, 2012).

22 María Puig de la Bellacasa, *Matters of Care.*

23 Alison Mountz, Anne Bonds, Becky Mansfield, Jenna Loyd, Jennifer Hyndman, Margaret Walton-Roberts, Ranu Basu et al., "For slow scholarship: A feminist politics of resistance through collective action in the neoliberal university," *ACME: An International Journal for Critical Geographies* 14. 4 (2015), 1235-1259.

24 Ibid.

25 Ibid.,1237

26 INCUBATOR Lab, *Life in the Soil*, 2017. Event website: https://lifeinthesoil2017.wordpress.com.

12 Ethical Tours & Hunting International

Christian Kuras

Image Credits

Christian Kuras, *Ethical Tours & Hunting International (ETHI)*, 2017. Concept website.

ETHI I home

www.ethi.hunting.com

ETHI

ETHICAL TOURS & HUNTING INT.

Hom

WHICH LARGE SPECIES WILL BECOME EXTINCT IN YOUR

Which animals will die out, and make it impossible to ever hun
At ETHI, we are dedicated to making it possible for you to clair
members of these species, before they are all gone. Sometime
ultimate in hunting: the last known surviving member of the sp

LATEST OFFERS ON LAST CHANCE SPECIMENS

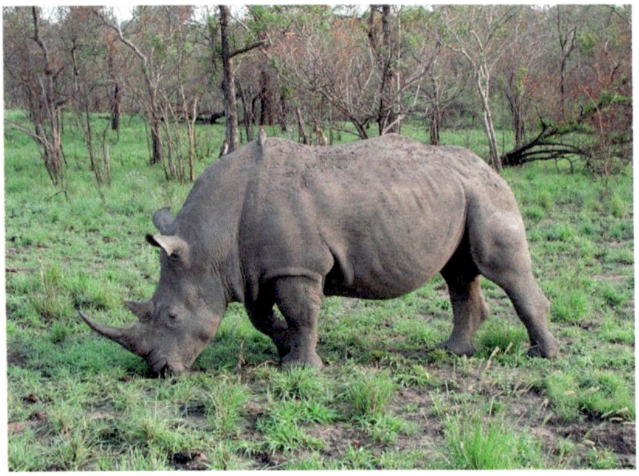

BLACK RHINO ADVENTURE

Be the one to take the last surviving wild Black Rhinoceros! Book now for our next tour: stalk and take the final specimen in the wild of this majestic beast, sadly driven to this point of extinction by irreversible climate loss. We take care of all arrangements and licencing. Click below to see more details.

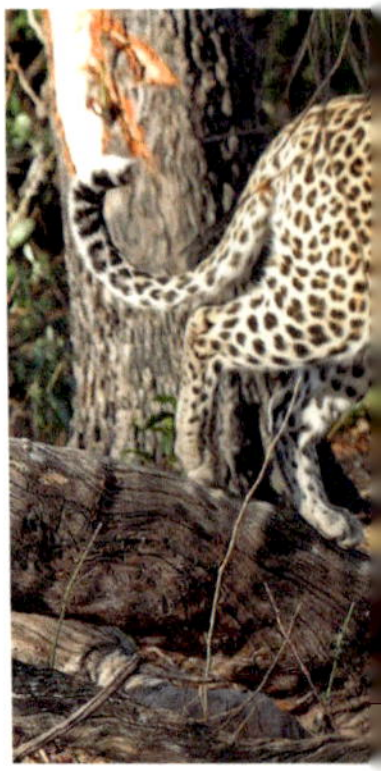

THE THREE LADI

A permit has ju
rights to three
of Sommerset
the park's pop
three adult fen
of reestablishi
extreme isolat
breeding popu

Google

out us Hunting Adventures Safari Tours Contact us

IE?

gain, for anyone?
the few remaining
ı even offer the

Find out more about upcoming tours and hunts

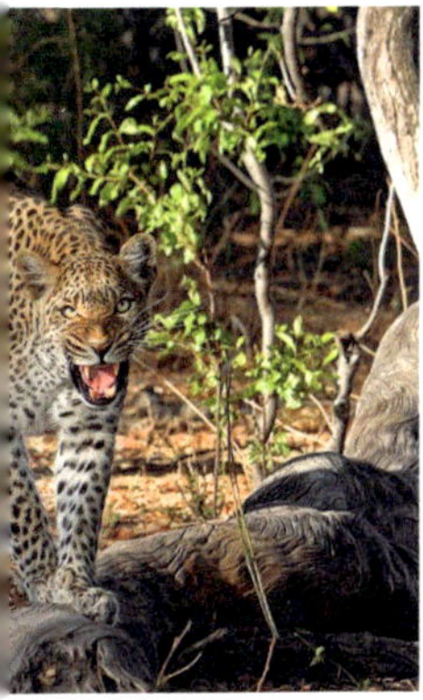

ne available for the
ıg wild cheetahs
e to poaching,
reduced to
th no possibility
pulation, and
other existing
he kill is now to

OUR ETHICS

We all know that the tragic loss of natural habitat has rendered many species untenable. But we have a way to mitigate the damage. The fee for your hunting experience will be directed towards relief from the effects of climate change on the host countries.

Scientists have known for decades that extinctions are a normal and natural part of the life cycle on earth. There have already been 5 mass extinctions before the one we are experiencing now. ETHI offers the opportunity of taking part in the Earth's natural evolution.

ALL INCLUSIVE ADVENTURES

After the thrill of the hunt, relax in our

13 Dynamic Stasis
A Perspective on Representation in Bioart

Marta de Menezes

The dichotomy between stasis and change can be used to help understand, and to some extent shed light on, some of the issues that are raised when one thinks about representation in art and biology. More importantly, an examination of this dichotomy can provide perspective on our reactions to bioart and to other related fields in the arts. Stasis and its opposite—movement—are, for me, key concepts that I will use to explain some of the reactions and experiences one feels when making and viewing art. I consider stasis to be linked more to traditional media, particularly to those aspects concerning representation; living materials and other life-like characteristics of matter and its manipulation in the arts are more directly linked to movement and a more energetic, dynamic, state of matter.

I have recently read a series of fantasy books that emphasize the dichotomy of stasis and change, in which some of the characters are able to shift space by synchronising themselves with the matter around them. It is a challenging idea, and I often wonder whether we invented the concept of stasis simply in order to allow us time to understand change, i.e. to understand movement and life. In physics classes during high school, pupils learn the crepuscular hypothesis, which states that matter is made of molecules and atoms that are always in a more or less excited state of being. According to this view, everything is made of moving particles and the spaces between them. This means that at an atomic level, everything is in constant movement, constantly subjected to change and transformation. Understanding this allows us to see stasis as a conceptual framework and helps us comprehend the complementary idea of perpetual motion—an idea we continuously struggle to understand. What does it mean to recognize that all matter is in a state of perpetual change and excitation? It feels unnatural; it escapes our control, something that we will always have issues dealing with.

Living Media

There are a few questions that are frequently asked in relation to my artwork and my approach to bioart. In relation to the use of a living medium, I am often asked about the reason for choosing such a medium and its relevance to the concept of the artwork: why is the manipulation of living material such a key factor in the artistic process I pursue? Interference with biological mechanisms is important for multiple aspects of my art practice (the conception, the research, and the production of the artwork). In fact, biological materials and their manipulation are also important after the completion of the artwork, at the time of its public exhibition—the time of contact with the viewer, the spectator, and the public. The work of art is only really complete when a spectator sees it and generates a response, and in such instances the use of living material can play a very important role in understanding and interpretation.

Key to understanding an artist's desire to manipulate life are two factors: first, the knowledge that something alive is under transformation, change, and always moving; and second, the concept that modification is itself something ubiquitous to art. Art is, and has always been, about life, independently of the matter that we use to make it. Using life itself, or a matter that expresses that quality (liveliness), gives the artwork an edge in relating to its core subject matter: life. This clearly brings us back to the ideas of stasis and movement, since whether something gives rise to the sense of movement (as life) or, by contrast, stasis, it has a strong impact on the choices an artist makes while creating art. Since life, by definition, is constantly transforming it is evident that using life as a medium will generate a response in the spectator closer to the living organisms to which the art piece refers. Biological media allows artists to manipulate dynamic and evolving systems in a way that cannot be reproduced in more traditional media. The two concepts (stasis and movement—or death and life or unchanging and dynamic) are equally important to the making of the artwork and to the perception/interpretation of the artwork. Thus, I feel that it is necessary to think about representation in bioart at those two different stages of any project: during its development and at the time of its exhibition. The way representation impacts those two stages is also, in some aspects, different.

Nigel Thrift, discussing his concept of non-representational theory, states:

> It would be possible to argue that human life is based on and in movement. Indeed, it might be argued that it is the human capacity for such complex movements and the accompanying evolution of movement as an enhanced attractor

> that has produced the reason for much of our rhizomatic, acentred brain.[1]

According to this way of thinking, there is a link between movement and life, hardwired into our brains and into our biological limitations or characteristics (depending on whether we adopt a more positive or restrictive way of seeing ourselves). This means that we perceive and respond to life, or lifelike stimuli, through a default mechanism. Simply put, movement signals a capacity for life while stasis means biological death. I believe that this is also the reason why we empathize with other organisms. It is the reason why we develop immediate connections to lifelike works of art and to living material, both when they are made (the artist is drawn to a lifelike material or a lifelike manipulation of material), and when they are experienced after their completion (lifelike artworks, or works that involve in their presentation make us pay attention). We recognize life, we develop empathy toward life, and it touches us, leading us to make a stronger effort to understand what is behind it. We want to understand the meaning of lifelike artworks and to develop a relationship with them. But this process is not exclusive to the final stage of an artwork, when it is publicly displayed, especially not for artists. This attraction also happens when artists are deciding which matter to use for concept development or in the production of an artwork. During the conception and production stages, the artist responds to something alive or lifelike. It is a type of response that is arguably different from what is elicited by an inanimate, or static looking object. Indeed, the use of living materials conveys complex and transformative concepts and elicits feelings that invite manipulation and an exploration of what it means to be living. The fact that the outcome cannot be fully predicted leads to the development of manipulative processes and techniques, influenced by the artist as well as by the material. While we roughly understand our fascination with movement and life, we also have a strong survival instinct to control it, and to make sure it is not a risk to our wellbeing.

Thrift also mentions in the introduction to his book that "every creature, as it 'issues forth' and trails behind, moves in its characteristic way."[2] In fact, all artists who work with biological or lifelike materials also have their characteristic ways of engaging with different directions in the larger field of bioart. Bioart needs to capture change, growth, and the constant dynamics of life in order to represent concepts that are also changing. As life itself is infinitely varied in its multiple forms, so are the approaches to it. Every artist has his or her own specific concepts to explore, using different materials and processes to do so. The artist, the maker, needs to choose the most adequate media to manipulate in order to convey, suggest, or trigger the intended meaning to the viewer, that which will then be perceived and interpreted singularly while in the presence of the work. Matter or media is to be understood here as the raw material: in the case of bioart it is usually whole organisms (like plants, bacteria, insects or mammals) or parts of organisms (such as cells or organs), or even smaller parts that can be synthetic (like molecules, which are the components of all living materials). Technology provides the tools to preserve, manipulate, and visualize these materials.

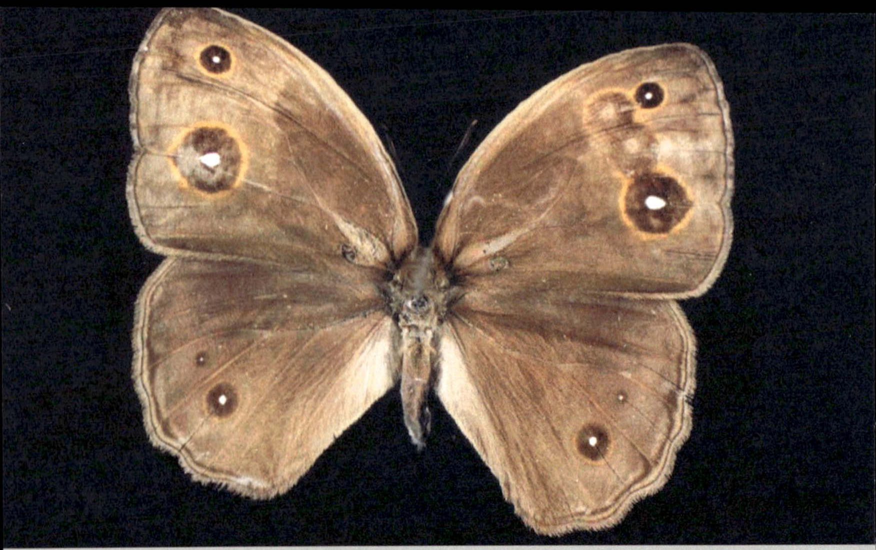

NATURE?

In *Nature?* I have created live butterflies with wing patterns that have been modified for artistic purposes. Such changes were achieved by interfering with the normal development of the wing, inducing the development of a new pattern never seen before in nature. The butterfly wings remain exclusively made of normal cells, without artificial pigments or scars, but designed by an artist. These wings are an example of something simultaneously natural and resulting from human intervention. The artistic intervention leaves the butterfly genes unchanged, thus the new patterns are not transmitted to the offspring of the modified butterflies. The new patterns have never before existed in nature, and will rapidly disappear from nature not to be seen again. These artworks literally live and die. They are an example of art with a lifespan—the lifespan of a butterfly. They are an example of something that is simultaneously art and life.

Developed at the laboratory of Professor Paul Brakefield, University of Leiden, Holland, with the scientists: A. Monteiro, M. Bax, K. Koops, R. Kooi and P. Brakefield.

Public Space
Between Representation and Presentation

In *The Emancipated Spectator*, Rancière states that a work is not an artwork until it reaches the public. Until then, it may be considered an art project, and it may still be shown or published before it is finished. However, once a piece is exhibited then it is an artwork (as it has reached the public), if not, it is in the process of becoming one. It is now very common to present the research before it reaches the final stage of production, often shown as work in progress as a strategy to better prepare for a more effective final artwork. We effectively bring the spectator into the creative field of making art, acknowledging the importance of the viewer in the making of meaning. And this is also why we bring it forward to the public sphere even before the artwork is finished, to take full advantage of the effect of the viewer as creator of meaning, to make the artwork the most effective trigger of meaning. This is not without complexities however, since the role of the spectator has historically been one of more passive engagement. Rancière describes it in the following way:

> I shall call it the paradox of the spectator—a paradox that is possibly more fundamental than the famous paradox of the actor. This paradox is easily formulated: there is no theatre without a spectator (if only a single, concealed spectator, as in the fictional performance of *Les Fils naturel* that gives rise to Diderot's *Entretiens*). But according to the accusers, being a spectator is a bad thing for two reasons. First, viewing is the opposite of knowing:

> the spectator is held before an appearance in a state of ignorance about the process of production of this appearance and about the reality it conceals. Second, it is the opposite of acting: the spectator remains immobile in her seat, passive. To be a spectator is to be separated from both the capacity to know and the power to act.[3]

Adapting this idea to the question of bioart, and not simply theater, we may be tempted to suggest that living art is not representation (it does not simply represent life or a concept related to it), and is instead a form of presentation (it uses life itself to trigger the interpretation of the perception of life by the observer). We might think of this as a process in which art made with living material, or with a lifelike appearance, is not representing something other and outside of reality. In bioart, life itself presents a concept that relates to the reality of the living material. If we understand the question in this way, we gain a different perspective on the strategies used by many bioart artists to interact with the public. Many bioart artworks have a strong interactive component, the most basic manifestation being the fact that it becomes fundamental to the piece that the viewer understands the living condition of the artwork. As we progress towards a more interactive society where everyone feels more and more entitled to an individual opinion, to a stance, and to an active position in society, this may even be something to expect from all cultural activities. The kind of (historical) spectator that Rancière describes no longer has a space in contemporary bioart, where meaning and the piece of art derive from the interpretation of the viewer and their emphatic connection to the work itself.

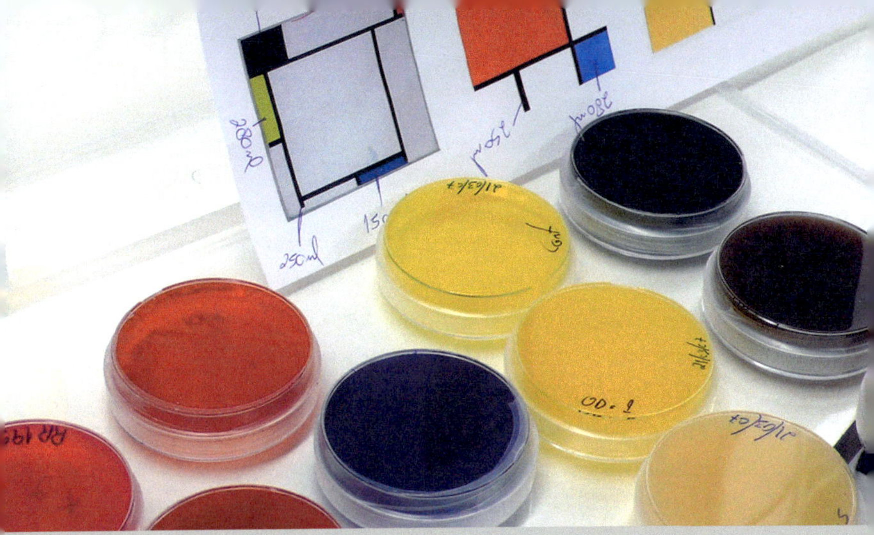

DECON: DECONSTRUCTION, DECONTAMINATION, DECOMPOSITION

Decon uses biotechnological materials to develop paintings that are literally alive but that deconstruct themselves while exhibited. Replicas of Piet Mondrian's geometric paintings are created using a bacterial support medium; the colors in the paintings are progressively degraded by *Pseudomonas putidabacteria*. The technology is based on the work of Dr. Lígia Martins at the Instituto de Biologia Química e Biológica in Lisbon. In her laboratory scientists research biological strategies to degrade highly pollutant textile dyes using bacteria harmless to human beings and to the environment. During the development of *Decon*, my collaborators and I researched the optimal conditions for influencing bacterial activity, adapting the color degradation rate to the environmental conditions of a museum. The objective was to achieve a slow decomposition of the images during the duration of their public exhibition.

Developed with help from Ligia Martins (MET Group, ITQB). With support from SOPHIED and Direcção Geral das Artes.

Science has been strategically promoting public engagement and fostering public understanding of science for many years. These strategies have become significantly more prominent in the last 20 years and have contributed to greater social support for science, as we see, for instance, with the growth of charities dedicated to sponsoring biomedical research. "Open days" have become commonplace in research institutes, where visitors can be involved in "hands on" experiments and discussions with scientists. But although these activities—which might include activities like DNA extraction or electrophoresis—contribute to greater awareness of science and participants become more knowledgeable about science, it is uncertain to what extent this actually allows society to influence the pursuit of science. While, on a large scale, policy makers and charities (a good example is the Bill and Melinda Gates Foundation) can influence the research effort by directing resources to specific issues, on a more human level, the interactions of scientists and the public appear to have little influence on the science. In this way, "open days" resemble a theater production, conveying scientific research to the masses. In this instance however, the spectator is a passive agent. Conversely, although it may sometimes seem that bioart is imbued with a similar relationship, it is actually not a practice where the viewer is dispossessed of knowledge or the ability to impact meaning. The bioart viewer is not passive. At the same time, it may not be as simple as stating that one is representational while the other is presentative. Instead, like with the question of life itself, which cannot be reasoned as being simply one or the other, it may be necessary to think of bioart as a combination of both. Each artwork has its own place, a position within a spectrum between presentation and representation, moving towards one extreme or the other as the viewer becomes more or less influential in the process of knowledge making. For thinkers like Rancière what is at stake in this relationship is the question of community:

> A true community is therefore one that does not tolerate theatrical mediation; one in which the measure that governs the community is directly incorporated into the living attitudes of its members.[4]

Because knowledge is often inseparable from acting, the idea of pure representation has limitations in terms of its capacity for transmitting knowledge. However, an artwork which is about actively manipulating life is embodied in knowledge derived from the process of its production, which also provokes the spectator to learn more. Bioart, consequently, cannot be purely representational in the way an image is or it risks losing a significant part of its impact. On the other hand, neither is bioart purely presentative, in the sense that life is not presenting itself in a simple way (there is an artist involved, afterall). Bioart, in my view, provokes the desire to learn by presenting a complex manipulation of living forms. By this I mean that the living cells, organisms, or molecules, and their manipulation, stand in place of an idea (or representation). It is an unmediated presentation that reaches the spectator in full force. It takes advantage of the living and the animation of life.

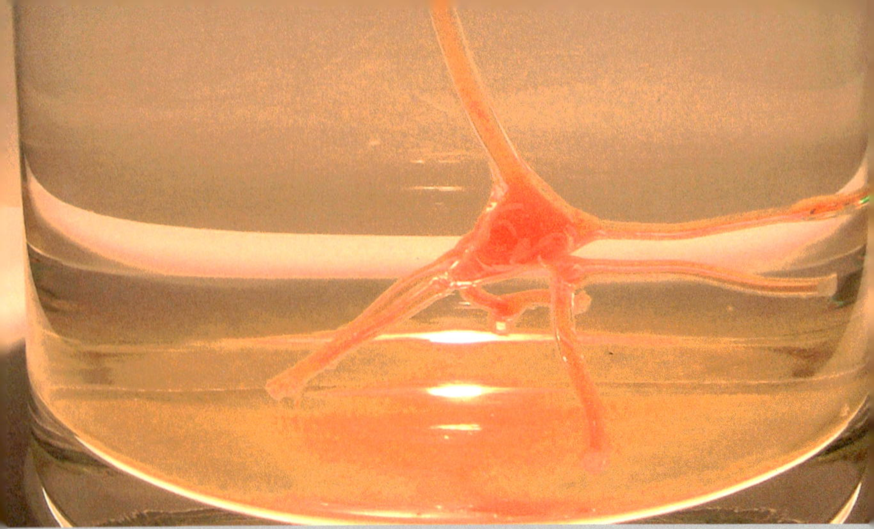

TREE OF KNOWLEDGE

Tree of Knowledge explores interactions between art and science by using novel cell imaging and tissue culture technologies in order to create living sculptures. When considering what would be the most adequate medium to represent the 3D structure of neurons, I concluded that I ought to use neurons themselves. By covering a scaffold with live neurons, or by filling glass tubes with these cells, one can achieve a representation of the delicate structure, maintaining the dynamic nature of a neuron: always changing, establishing new connections, eliminating old ones, growing, living. It is important to note that this strategy is not an attempt to represent reality as accurately as possible, but simply to explore the material that seems more appropriate for the representation of such a living object.

Developed at SymbioticA, University of Western Australia, in collaboration with scientists from Dr. Giles Plant's group and artists from SymbioticA (Oron Catts, Ionat Zurr). With the help of Dr. Miguel Vaz Afonso (Max Plank Institute Munich).

Art and Biology
Beyond Representation and Non-representation

One may argue that bioart is, above all else, an art of non-representation, in the sense that Thrift defines non-representation as something experimental and in search of a different kind of concept.[5] As Latour states, "after all, no battle has ever been won without resorting to new combinations and surprising events."[6] In particular, bioart pulls the energy from the performing arts and injects it into the social sciences in order to make it easier to "crawl out to the edge of the conceptual" (as Helen Vendler might say).[7] Or as Sandra Kemp put it: "To see what will happen. To let the event sing to you."[8]

Thrift explains that this style of (performative) thought is about the "how" and not the "what": "non-representational theory is an approach to understanding the world in terms of effectivity rather than representation; not the what but the how."[9] And although it seems tempting to place bioart in this non-representational field it is not quite as simple as that. The "how" may be a fascinating part of works involving biology, since it helps to understand the science and technology involved in making the works, but that is hardly ever the sole reason behind the making of the artwork itself. The "what" is relevant for a bioart artwork because that is what generates curiosity, and affect. It is in the final manifestation of the piece that the work achieves its purpose: where the artwork becomes an artwork. An audience must be able to see beyond the process (the "how") in order to generate meaning. In the end, we must not discard the representational factor in bioart.

The role of the spectator in bioart seems to take on an importance that we often do not associate with painting or sculpture. Instead, a bioart spectator seems to associate more with performance and theater and in particular with a "better" theater. Rancière describes such a form of theater as follows:

> We therefore need a different theatre, a theatre without spectators: not a theatre played out in front of empty seats, but a theatre where the passive optical relationship implied by the very term is subjected to a different relationship—that implied by another word, one which refers to what is produced on the stage: drama. Drama means action. Theatre is the place where an action is taken to its conclusion by bodies in motion in front of living bodies that are to be mobilized. The latter might have relinquished their power. But this power is revived, reactivated in the performance of the former, in the intelligence which constructs that performance, in the energy that it generates.[10]

This is a very powerful quotation and I believe we can apply Rancière's words directly to questions of art and biology. Artists working with living materials provide more than the simple and passive observation of a phenomena or even a demonstration of a technique. We offer action, and we offer the idea that that action is shared between the artist and the audience. By experiencing the artwork, one is not a passive spectator, but a fully responsible actuating participant. The spectator completes the work by viewing it, experiencing it, and being pulled into the knowledge they acquire from the experience. This is the drama and the tension of bioart. I believe that this involvement is achieved through a form of empathy that is provoked in the spectator, an empathy that makes them identify with the living Other in the artwork, compelling them to exchange the position of a passive spectator for that of an active experimenter or even a scientific researcher, making choices, walking in someone else's shoes, and facing the dilemmas that the artwork poses. Sometimes I wonder if this connection to performance, to theater, is the reason why so many artists with backgrounds in the performance arts decide to work with biological material.

> The paradox of the spectator pertains to the curious device that adopts Plato's prohibition of the theatre for theatre. Accordingly, it is these principles that should be re-examined today. Or rather, it is the network of presuppositions, the set of equivalences and oppositions, that underpin their possibility: equivalences between theatrical audience and community, gaze and passivity, exteriority and separation, mediation and simulacrum; oppositions between the collective and the individual, the image and the living reality, activity and passivity, self-ownership and alienation.[11]

In a way, Rancière's word describe one of the most important social functions of art, and in particular, art and biology. The objective is to make a connection between places that do not usually live together: exteriority and separation; mediation and simulacrum; collectivity and individuality; the image and living reality; activity and passivity; self ownership and alienation. It is an objective very dear to art, and one that is very strong with those artists working at the boundary between art and biology. What is important is not only the manipulation of life to convey meaning, but also the process of allowing an audience to generate meaning. This is the space where the spectator becomes a true part of the creative process.

> It is in this power of associating and dissociating that the emancipation of the spectator consists—that is to say, the emancipation of each of us as spectator. Being a spectator is not some passive condition that we should transform into activity. It is our normal situation. We also learn and teach, act and know, as spectators who all the time link what we see to what we have seen and said, done and dreamed. There is no more a privileged form than there is a privileged starting point. Everywhere there are starting points, intersections and junctions that enable us to learn something new if we refuse, firstly, radical distance, secondly the distribution of roles, and thirdly the boundaries between territories. ... We have to recognize the knowledge at work in the ignoramus and the activity peculiar to the spectator. Every spectator is already an actor in her story; every actor, every man of action, is the spectator of the same story.[12]

Bioart, with its use of scientific content that is still novel in the arts, brings a new problem to the already contested distinctions between subject and object in art practice. Art and life were historically separate and easy to distinguish. In the past, when someone visited a museum to see an artwork, however radical the work itself may be, one expected to find a representation of the living and not life itself on display. Not anymore. Nowadays, life itself can be found on display in art galleries and museums. And this is not just "life" in self-evident forms but a whole array of possibilities, including life cycles, or "units" of life (organisms, cells, or molecules) being modified, growing, and certainly dying—as death is an inescapable characteristic of what is alive. The experience is no longer restricted to a representation of ourselves and our living components. Instead our biological identity can itself be found on display, real, and alive! Importantly, all the while it is reproducing and dying, continuing to raise new question about the limits of the work.

This is a new form of realism, a realism that is not only a critique, but also a tangible reality, a visceral reality, and one that is necessary in order to force a commitment from the spectator to become a participant. In this way, the spectator becomes an active individual and therefore part of the community.

Notes

1 Nigel Thrift, *Non-representational theory – Space/politics/affect* (London: Routeledge, 2008), 5.
2 Ibid.
3 Jacques Rancière, *The Emancipated Spectator*, trans. Gregory Elliot (London: Verso, 2009), 2.
4 Rancière, 2-3.
5 Thrift, 5-6.
6 Bruno Latour, *Reassembling the Social: An Introduction to Actor-Network Theory* (Oxford: Oxford University Press, 2005), 252.

7 Helen Vendler, *The Breaking of Style* (Cambridge: Harvard University Press, 2005), 79.

8 Thrift, 12.

9 Sandra Kemp, "Reading difficulties" in ed. Patrick Campbell, *Analysing Performance: A Critical Reader* (Manchester: Manchester University Press, 1996).

10 Rancière, 7.

11 Rancière, 35.

12 Rancière, 36.

14 Unlikely Conspiracies of Biologic Thought

Kira O'Reilly

Taxi: Feeding Cells

It was around Christmas, during the height of the southern hemisphere summer, and hot. The cell cultures needed to be checked, their pink nutrient media was most likely some concerning shade of yellowish indicating depletion of substances essential to healthy metabolism, cell division, population and therefore maintenance.

Following the initial euphoria of discovery and initiation, there followed a weird depression that permeated during the secondary phase of their adjustments to working with life: moving from technologies of mechanical and digital reproduction into biological reproduction and re-presentation, from response to responsibility and care.

Inside the kitchen of the dark and shuttered house, away from the brilliant sun they admitted to their lethargy and depression.

They were handling and dealing with the see-saw of learning the rudimentary protocols of working with life, whilst seeking the base from which to form new ethical and caring relations with these minuscule, partial and extended forms. They wanted to ignore their charges: artstuff and processes.

"Let's go together and get a taxi there," she said, exclaiming something about Walter Benjamin and taxis. I've never been able to discover the provenance of the reference, or if it even it exists—and perhaps it is a mutation in my memory, a glitch of recollection. But a taxi was called and into it they went, travelling across the city to the laboratory.

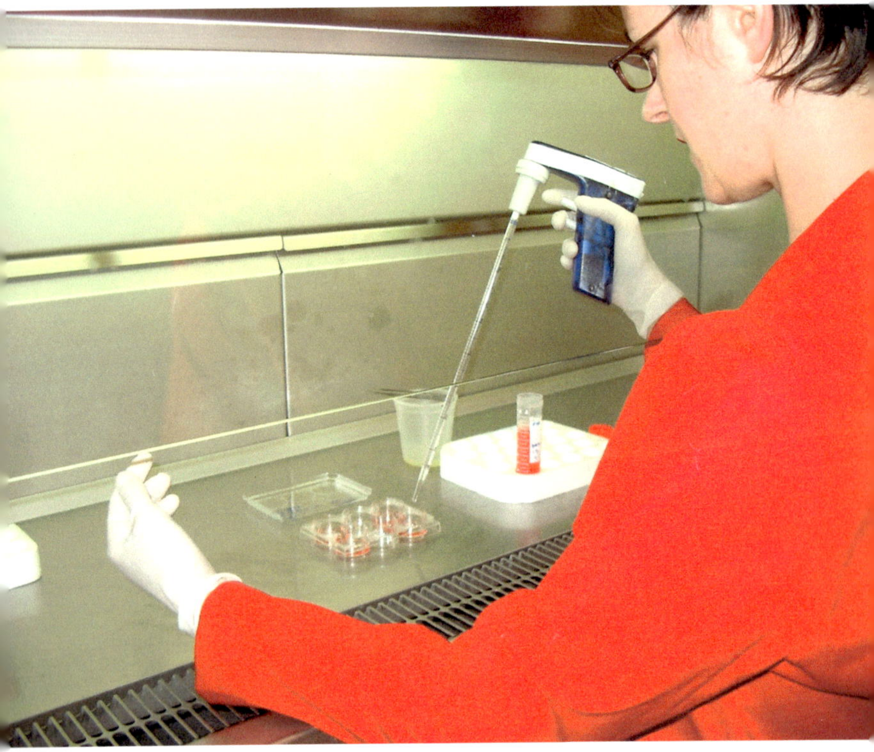

Kira O'Reilly, *Red Lab Coat Series*, 2004.
Color photograph, photo credit: Jennifer Willet and Jason Knight.

Being taxied as if in suspension, encapsulated, transported whilst the sky burned in a brilliant azure above the bare concrete and parklands of the city.

They found their way to where their cellular charges were, media was aspirated and new, fresh, vital liquid replaced.

Naked in the Lab: Intimate Proximities

During a residency at SymbioticA in 2004 I had wanted to bring languages, methods and practices from art making processes into the scientific environment I found myself in. The actual tissue culture laboratory was used by different researchers—those of the life science and those from the arts (from SymbioticA) alike. I learnt from artists such as Phil Ross, that a laboratory was as much constructed by its own culture—the rites, rituals and behaviors of the specific place—as by the architecture, kit and research.

Coming from a background of performance art, I understood this implicitly, and how context is material and as material as any other component.

One of my primary modes to investigate a site was via physically exploring it: inserting, installing, and discovering how my body might occupy it in normal and unusual ways. Turning bodily orientation upside down, back to front, seeking out odd relationships between my body and its environs so as to reveal how bodies are constructed and disciplined, in the Foucauldian sense.

To do this naked would be a typical mode and one that would add another dimension to what the human bodies in a laboratory space are doing. What are they? How do we understand them? As soon as individuals divest themselves of clothing in nonconventional contexts, all kinds of other readings run havoc. The body, marked by gender and other signifiers of identity, or bodies regognized, mis-signified and otherwise, come into play.

I never did get to make these investigations, quite understandably it was considered to be too much, and

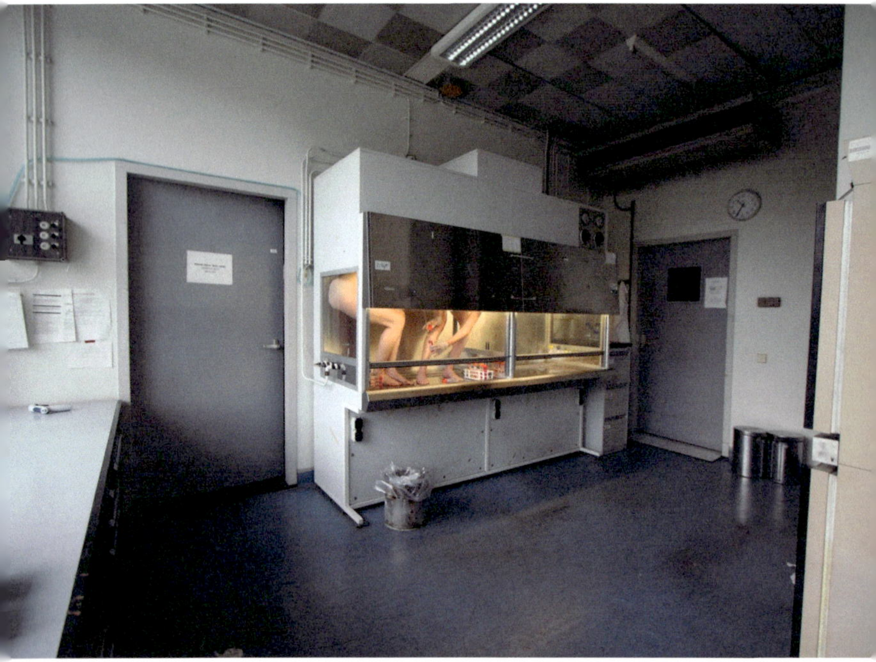

Kira O'Reilly & Jennifer Willet,
Untitled (Hamster Ovaries Protocol), 2008.
The Art and Genomics Centre, University of Leiden,
photo credit: Rune Peitersen.

perhaps off-putting to the greater research community and possibly unhelpful towards maintaining and cultivating thoughtful and sensitive bridges across and between disciplines. However the idea and its merit persisted.

Years later, Jennifer Willet sought me out for a meeting in London with a proposal for a collaboration. She had found her own way to working with similar methods of situating the body and specifically her own body in imitate proximities to the organic and other living other bodies she was working with—in laboratories or out in the field. She had already made a photographic series with permissions from the University of Leiden's Art and Genomic Centre, taking swabs from sites of her nude body. In these work the discursive body was not presented as being solely human in material or ethical terms, but instead a shifting ground of relations. Her proposal was that we join our discreet artist practices and together make a series of photographic images, collaborating on situating our bodies in relation to the laboratory as a site and, crucially, implicating our bodily selves into the operations of the life sciences and its technologies.

Gentleman Scientist, Interloper, Pedagogue

She was always a gentleman scientist at some level, a Woolf-like Orlando. "I am an interloper," she said as she introduced with a charming smile from behind her stylish but careful spectacles. There was always a little hesitation that she allowed her addressee, as if to say: "Do you believe me? Do you trust me?"

The gentle man assumed discretion and travelled through times, conventions, bureaucracies. He had learned the language of institutional lore, the careful negotiations of committees and processual protocols and deployed these accumulated fluencies in her pursual of his insidiousness of biologic logic and illogics.

She was quite the pedagogue, this wolf-like Orlando, teaching with intricate fastidiousness in the ad hoc laboratory made in the prefab school behind the parking lot. She did not exaggerate when he described the puddles and clamber through a wire fence to get to the incubating art lab.

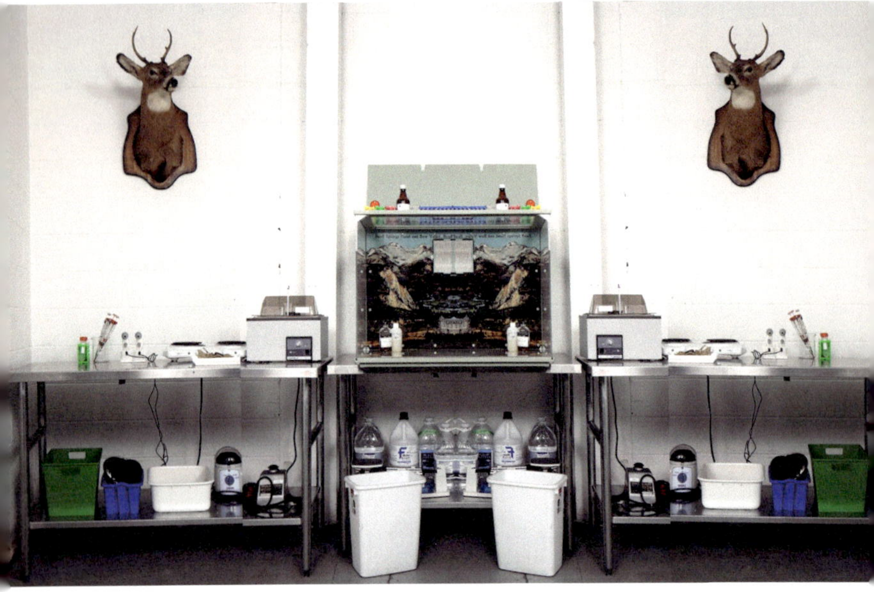

Jennifer Willet, *INCUBATOR Lab Wall Mural*, 2013. Installation at the Ontario Science Centre, Toronto, Canada, photo credit: Arturo Herrera.

Decoration was an articulation of elaborating on knowledges, reshuffling and examining into a celebratory collusion of craft and crafting, biotechnologies and artful makings in the pedagogue's laboratory. The viewing rake of the anatomy theatre had been reimagined through layers of transparency, glass, screens, and streaming, everything shall be performed, he exclaimed, with the serious devotion of all play making. Nothing was done without humor, explanation or great trust and belief in his students, all drawn from the local town.

The exterior surfaces of the tissue culture laminar flow hood danced with designs from William Morris-like block wallpaper etched deep into their steel exteriors. The interior of the dark cell culture incubator was marbled into swirls with deep antibacterial copper and Verdigris.

The paper disposable laboratory coats collars stood to attention in Rembrandt-like stiffness, extravagant sleeves caught into wrist holding gathers. Sometimes students would catch themselves covered in such crispness, their attentiveness enhanced and nuanced accordingly as she guided their profound contemplations of working with the biological and ecological entwining of life and lives.

Grids, Frames, Unlikely Conspiracies of Biologic Thoughts

Unruly taxonomies
Re-arranging disorders
De-spatialized bodies
Grids that warp taxonomic time
Euclidian crumples and vanishing points

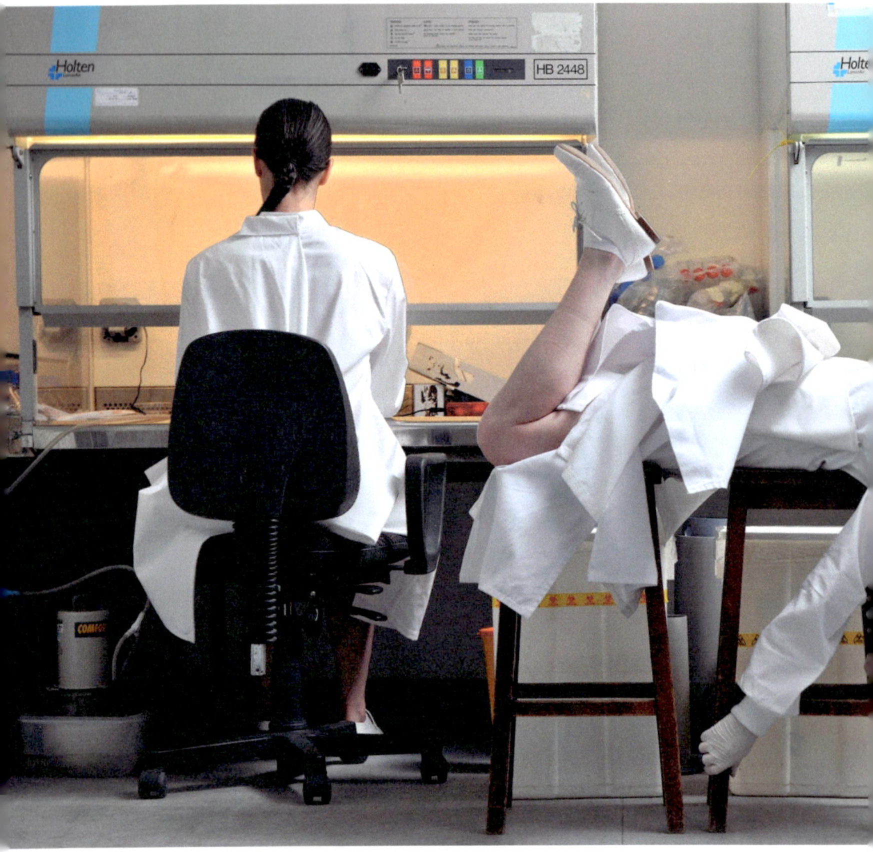

Kira O'Reilly & Jennifer Willet,
Refolding (Laboratory Architectures Twins), 2010.
School of Biosciences, University of Birmingham,
photo credit: Hugo Glendinning,
laboratory coat construction: Shanti Freed.

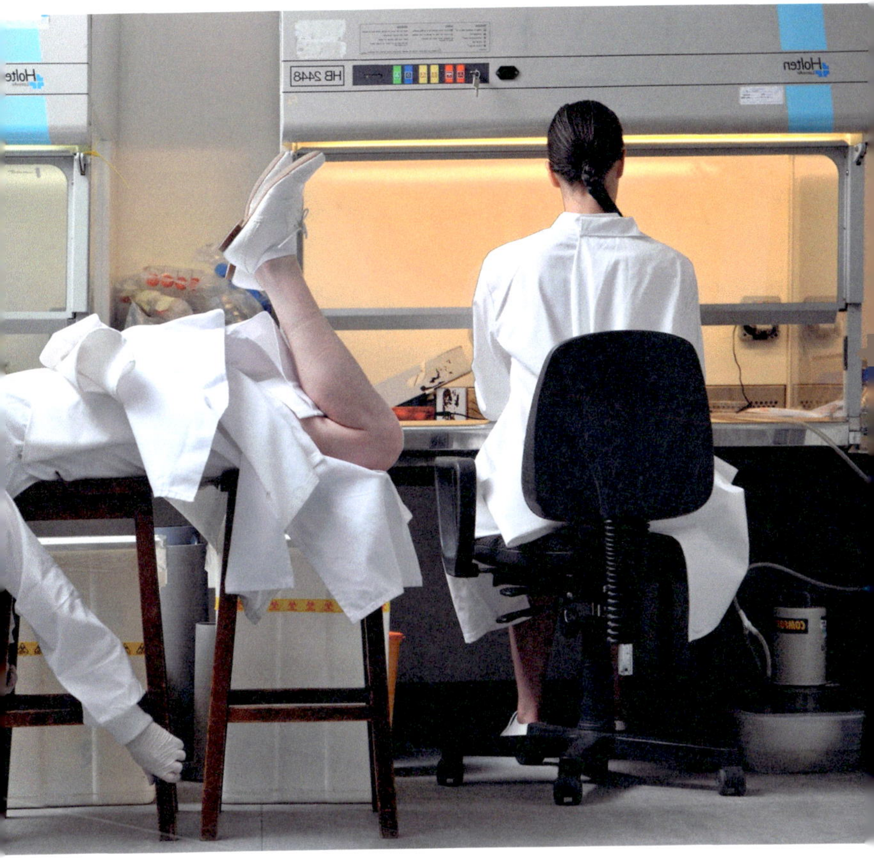

They spoke to one another of *The Draughtman's Contract* (1982), *A Zed and Two Noughts* (1985), *Drowning by Numbers* (1988), all films by Peter Greenway and of his methods of organizing perspectives, ordering durations and plotting narratives with perspective enabling grids and categorizing mechanisms.

Multiple laboratory coats were sewn and rearranged into two otherly coats, taking inspiration from the excessive seventeenth century frock coats in *The Draughtsman's Contract*, kimono-like foldings and stiff obi wrappings. Protean pockets replicated, diploid rows of poppers sprung up, insect wing like protrusions and layered starched layering were structured into new architectures that allowed them to merge, mutate and meld into monster bodies within the photographic frames.

At the biosciences school they consulted with the Chinese Emperor in their disorder of the arrangements of bodies and bits of bodies in the pictorial field, perhaps in the hope of warping scoping drives. The Dean worried that the dripping squid bought from the market that morning would tell of anti-ethical practices, unhappy researches and perhaps errant researches.

They played, moved, and re-arranged their bodies peculiarly, seeking unlikely conspiracies of biologic thought and ecological re-orderings whilst moss mossed and fat, black flies in modernist wire cubes resisted photographic capture.

Environmentally Menopausal

I have been thinking about how to reframe menopause and to become a pirate.

> When I was a child, the only thing I wanted to be was a pirate. Because I wasn't a stupid child I knew that I couldn't.
>
> — Kathy Acker

Menopause is turbulent, one needs sea legs that are study and can move in accordance with changing weather, fortunes and seas. One is literally out at sea, uncharted sea over which the normal starry constellations are absent. There are starts but

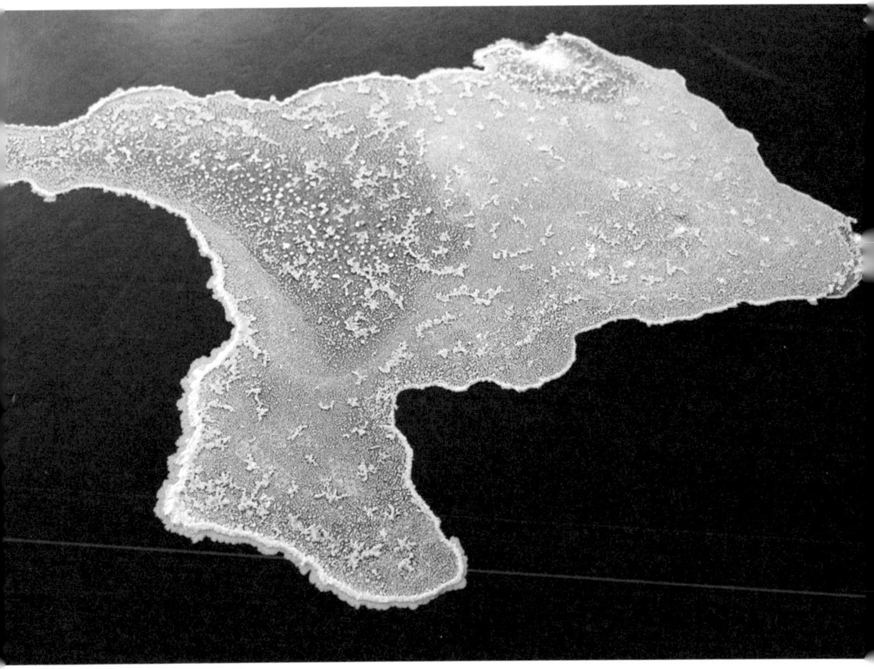

Kira O'Reilly & Flora Wellesley Wesley, *Salt Drawing*, 2015.[1]

they are arranged into hitherto unrecognizable orientations, the cardinal points have all shifted, perhaps the poles have moved to the equator and Lapland has come to me.

Menopause is a series of transitionary states, during which hormones soak, saturate and abate in exotic tides. One's very self is up for grabs and there is no clear sense of anything. This pervasive non-sense is rich and strange ground in which things do not exactly grow but emerge.

In our endocrine altered environments, insides and outsides are subtly permeable. Phytohormones and animal hormones, biochemical molecules that are hormone like, that masquerade and alter, all travel across and through bodies of plants. Animals in ecological relations as bodies are processes and processes are transformations of substances and things. Menopause is environmental.

Sweat Meteorology

Horses sweat
Men perspire
Women glow

I did not mind the arrival of the sudden night sweats that flashed like floods drenching me, despite their disruptive drama. Rather, I enjoyed their excessiveness and sense of abandon. I did not mind being sodden, somehow it was satisfying, intense and tropical.

Perhaps I found myself unbothered because I was used to sweating like a horse.

There was a moment some time prior to the onset of hormonal deluges when I announced that I "sweated like a man" when I trained. There was nothing restrained or glowing about my salty deluges. I poured unrelentingly, until sodden. My face went red. My hair all over the place. I precipitated, misted and formed clouds.

Sweat Protocol (i)
- Be menopausal
- Go to bed
- Throw off covers

Sweat Protocol (ii)
- Hold a copper pipe closely in the heat of the day under the hot sun
- Glisten

Sweat Protocol (iii)
- Hold an iron kettlebell
- Perform swings energetically
- Sweat
- Cause it to rust

Sweat Protocol (iv)
- Wear white cotton mutated lab coat to the gym at work
- Run and/or row into a heavy sweat
- Use coat to absorb sweat
- Bury the sweaty coat in a wild area in the hope that sympathetic microbes will grow on the sweaty cotton

Trust Me

Do you trust me?
Why do you trust me?
Do I trust myself?

Questions and tensions that derive from institutional rules, their observation and abeyance have permeated our collaborations as artists and as friends, sensing and finding the edges of what is possible, permissible and desirable even.

Jennifer Willet has an exemplary track record of working with large institutional bodies and their organs of funding and research, she has shouldered significant responsibilities in order to work with these operations from within in order to fashion art works and enable education that is directly engaged and deliberately cares about contemporary issues and conditions.

My pathways have tended towards the idiosyncratic, private, domestic, often freelance, and outside of (although by no means always) institutional parameters. Informed to some extent at least by the DIY assertions of post-punk subcultures and subsequently by certain feminist art practices to challenge operations of power via expertise and license, I've sought to find tactics and strategies. Our lives have taken very different shapes and pathways but these tender tensions we ruefully pull on as they are made of tissues and threads that enable thoughtful unpicking, examining within our individual and collaborative works.

Increasingly we afford one another spaces for multiple perspectives to be experiences and for the potentials of these situations to be examined. There is a currency of friendship and a certain irreverence, mischief is welcomed

Jennifer Willet & Kira O'Reilly, *Be-Wildering*, 2017.
Color photograph, photo credit: Bas de Brouwer.

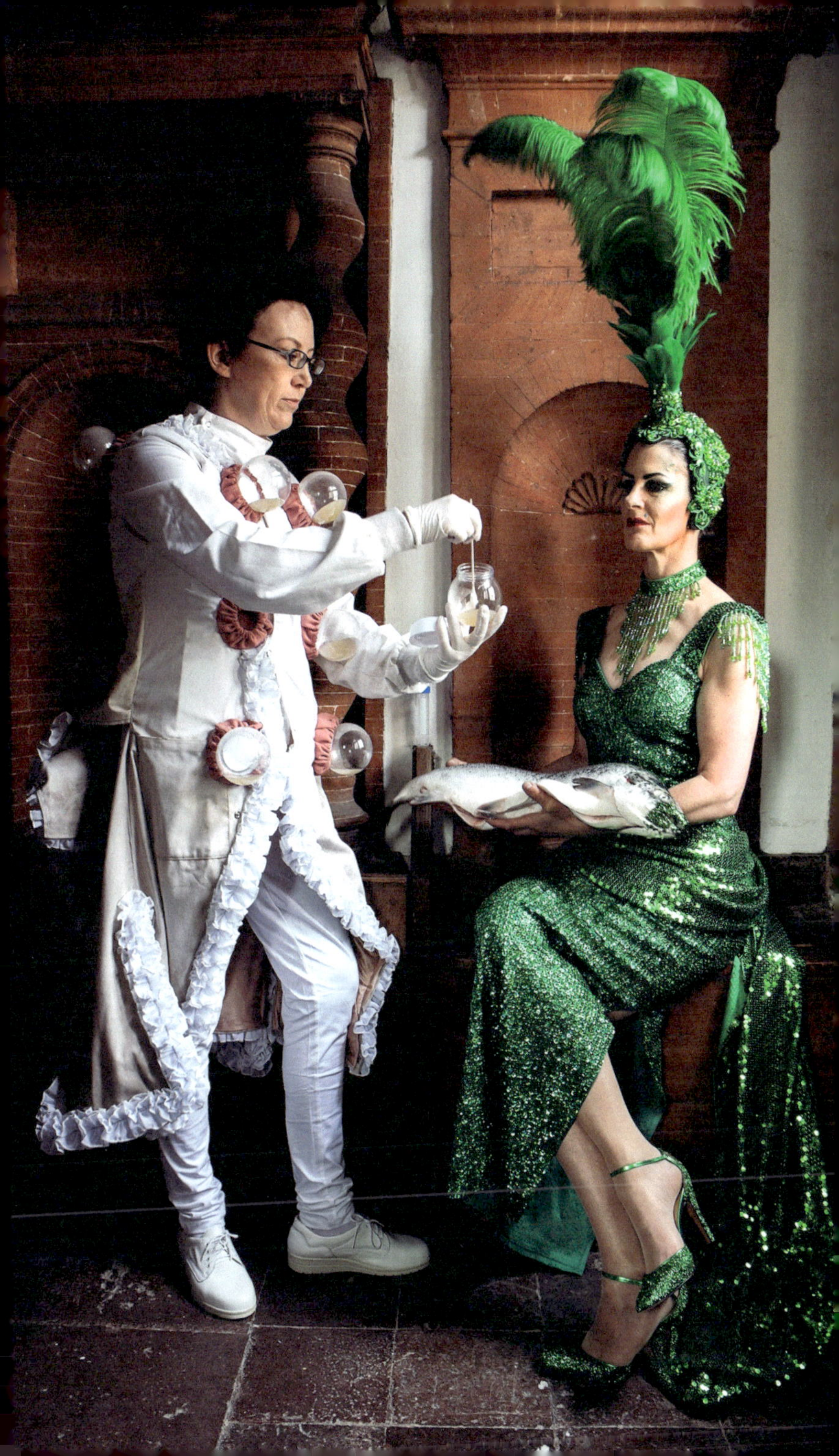

into these currents, humour and willful play, silliness even. What is at play is perhaps is a daunting dismantling of authorativeness of institutional knowledges, in order to locate, and to lovingly cultivate, deeply living with, thinking with, and being troubled with multiplicity of life in all its registers.

The Salmon of Knowledge

In the water of the artificial lake she stood, the cold water coming up to the middle of her thighs; she has slowed down her breathing to long, slow cycles of inhalation, pause, exhalation, pause so as to manage the effects of the cold. The backdrop was the extended concrete dam of the hydro-electric plant through which the lakes waters had been released only a week previously to fill the stoney lake bed. With the flood of water the concrete salmon stairs had also filled to a create watery pathway for the wild and muscular Baltic fish with their wide and powerful tails, to make their way upriver to spawn.

Her stockinged feet began to become numb as she stood on slippery rocks, dressed in the unlikely and surprising attire of an emerald green sequinned dress with elaborate and heavy make-up, topped by a sequinned and feathered headdress.

People began to appear on the shore, dressed in warm coats and hats to keep the cold out, through the damp, dank day they stared out at her.

After some minutes she began to move towards them slowly and carefully, taking care not to fall, the slippery unevenness underfoot and the soaking heaviness of her saturated gown made her ungainly. She allowed herself to sway, slide, and wobble, the delicate ostrich feathers of her headdress elongating her length into dangerous angles.

A pair of green glittery shoes sat like gems on the shore. On reaching it she sat on a large rock, placed them on her shivering feet and fastened them, amazed that the tiny buckle found its way to close with such ease.

Nestled in the rocks was the glistening dead body of a heavy salmon that had been pulled from the waters of a Norwegian fish farm to be sold. She gently sprinkled emerald

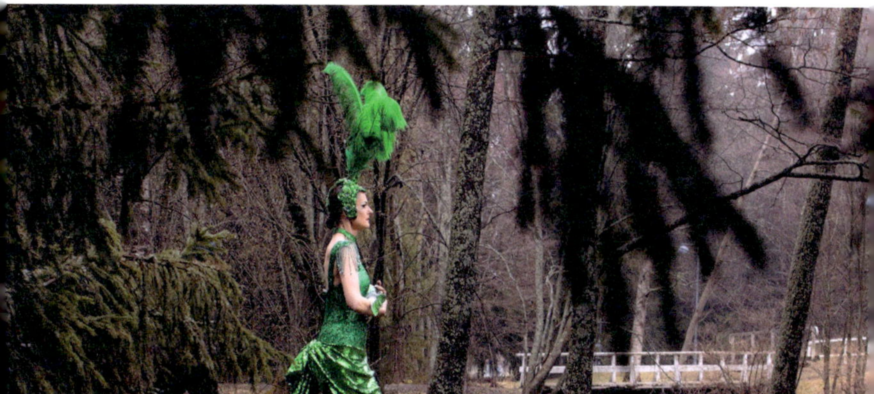

Kira O'Reilly, *Be-Wilderment (Nature Drag)*, 2017. Performance photograph, photo credit: Mika Friman.[2]

green glitter onto its scales, marvelling at its gorgeousness and the effect of the scattering of the plastic microparticles onto the iridescent skin. It had its own cultivated charisma, an aestheticised organism firmly caught into nets of human preoccupations with food growth, market concerns, and rationalizations, most likely peppered with antibiotics to foil the opportunistic parasites that thrived in the farms. A Postnatural Salmon of Knowledge

Only a week earlier she had stood in the anatomy theater of the Waag with Jennifer, dragged up in the very same outfit, where she had poured a large bowl of identical glitter onto a mound of earth, feeling the gaze of everyone in the room enraptured by the dazzling cascade of light, fascinated by the troubling, strange and unnatural mixing of the glitter into the earth. There had been a farmed salmon there as well, also a postnatural emblem that she had placed onto the earthy sparkly mix and seasoned with glitter inserted the green glinting particles into its cut interior and mouth.

But these salmon tropes were real bodies, once alive with all the liveliness that permeates our radiant worlds.

Scooping the salmon into her arms she stood up, its skin was a little slimy, its body was cold, and she had to be careful to carry its weight so that it didn't slip and flop to the floor—a cold fish indeed. She gazed at it and at the watching crowd and began to move, walking through the surrounding tree filled parklands towards the museum.

She entered into the warmth of the museum where the sodden long folds of her dress dragged like a fish tail on the slate floor leaving a wet trail. Unearthly.

She moved from art work to art work slowly and with deliberate care, attentive to the works as if each one was calling out its own implicit requirements which only she could hear and thus convey to the dead salmon. It was as if each work was a milieu or world though with to move and in which to listen—deeply, occasionally she would whisper something to the salmon very quietly under her breath. She felt the works and herself re-combine into new or other otherings as she visited each one, strange splicings and unexpected activations.

Eventually she completed her tour, ascending the museum stair case, her sequined scale-like tail still wet and heavy. At the top of the stairs she paused, and went with the fish to stand beside and gaze into the painting *Ideal Landscape*, 1860, a window into a nature framed and caught by the 19th century eye of Werner Holmberg (1830-1860) the great Finnish landscape painter.

Later, she found, nestled in the crevices between the sequins of the gown, the bright salmon's clear scales.

Notes

1 This collaboration was curated for *Prologue*, a two week artist's lab, hosted by PanicLab in association with Artsadmin. Three pairings of performance artists were invited to explore and develop new ideas: Joseph Mercier/ Jordan Lennie & Mark Ravenhill, Jamila Johnson Small & Peggy Shaw and Kira O'Reilly & Flora Wellesley Wesley. *Prologue* hopes to foster a unique intergenerational and interdisciplinary exchange, giving way to the potential for new artistic projects and collaborations.

2 This work *Be-wilderment (Nature Drag)*, 2017, was made for *SPLICE, Re-Examining Nature*, an international art exhibition curated by Nina Czegledy in collaboration with the Bioartsociety and produced in partnership with the

Oulu Museum of Art, Finland. It drew on many sources including: *How to Explain Pictures to a Dead Hare* (1965) by Joseph Beuys, *Beautiful People* (1987), directed by David Wojnarowicz, and the *Salmon of Knowledge* from the Irish myths of Fionn mac Cumhaill. It was made a week after *Bewildered*, 2017, a collaboration with Jennifer Willet as part of the exhibition *Trust Me I'm an Artist.*

15 There is no such thing as gradual change

Christian Kuras

There is no such thing as gradual change. There is no such thing as gradual change. There is no such thing as gradual change. There is no such thing as gradual change. There is no such thing as gradual change. There is no such thing as gradual change. There is no such thing as gradual change. There is no such thing as gradual change. There is no such thing as gradual change. There is no such thing as gradual change. There is no such thing as gradual change. There is no such thing as gradual change. There is no suddenly all is gradual change. There is no such thing as gradual change. There is no such thing as gradual change. There is no such thing as gradual change. There is no such thing as gradual change. There is no sudden all as gradual change. There is no such thing as gradual change. There is no such thing as gradual change. There is no such thing as gradual change. There is no such thing as gradual change. There is no sudden all as gradual change. There is no sudden as gradual change. There is no sudden all as gradual change. There is no sudden all as gradual change. There is no sudden all as gradual change. There is no sudden all as gradual change. There is no sudden all as gradual change. There is no sudden all as gradual change. There is no sudden all as gradual change. There is no sudden all as gradual change. There is no sudden all as gradual change. There is no sudden all gradual change. There is no sudden all as gradual change. There is no sudden all as gradual change. There is no sudden all as gradual change. There is no sudden all as gradual change. There is no sudden all as gradual change. There is no sudden all as gradual sudden. There is no sudden all as gradual sudden. There is no sudden all as gradual sudden. There is no sudden all as gradual sudden. There is no sudden all as gradual sudden. There is no sudden all as gradual sudden. There is no sudden all as gradual sudden. There is no sudden all as gradual sudden. There is no sudden all as gradual sudden. There is no sudden all as gradual sudden. There is no sudden all as gradual sudden. There is no sudden all as gradual sudden. There is no sudden all as gradual sudden. There is no sudden all as gradual sudden. There is no sudden all as gradual sudden. There is no sudden all as gradual sudden. There is no sudden all as gradual sudden. There is no sudden all as gradual sudden. There is no sudden all as gradual sudden. There is no sudden change gradual sudden. All change is no sudden change is sudden. All change is so sudden change is sudden. All change is no sudden change is sudden. All change is so sudden change is sudden. All change is so sudden change is sudden. All change is so sudden change is sudden. All change is so so sudden change is sudden. All change is sudden change is sudden. Change is always so sudden change is sudden. All change is so sudden change is sudden. All change is so sudden change is sudden. All change is so sudden change is sudden. All change is so sudden change is sudden. All change is so sudden change is sudden. All change is so sudden change is sudden. All change is so sudden change is sudden. All change is sudden change is sudden. All change is sudden change is sudden. All change is so sudden change is sudden. All change is so sudden change is sudden. All change is so sudden change is sudden. All change is so sudden change is sudden. All change is so sudden change is sudden. All change is so sudden change is sudden. All change is so sudden change is sudden. All change is so sudden change is sudden. All change is so sudden change is sudden. All change is so sudden change is sudden. All change is so sudden change is sudden. All change is so sudden change is sudden. All change is so sudden change is sudden. All change is so sudden change is sudden. All change is sudden change is sudden change is sudden. All change is no sudden change is sudden change is sudden change is sudden change is sudden change is sudden. All change is sudden. All change is sudden.

There is no such thing as gradual change, or even the experience of gradual change.

All change is sudden. Every instant we're going over a cliff edge into the void. Time passing is how we explain this continual and ceaseless change. In fact there is no such thing as time, there is only change.

All change is expressed as ruptures of form, or rearrangements of form. Even something that looks still, and appears the same from one moment to the next, is constantly changing, and is not the same thing for any length of time. Its true constitution is change.

There are no things, only change. There is only change, ever. Change is unchanging.

This is why change is actually a silly and absurd idea. In fact there is no such thing as change, there is only the illusion called time. An investment in the idea of change is an ill-judged investment. (Because change implies that there are things that are subject to change, when, in fact, things reconstitute themselves every moment in the memory of what they were, but share only a trace morphology with their precedents. There are no things.)

An actual change would be stasis. That is really the end.

A One-Step Guide to Experiencing Change

1. There should be a feeling of something moving or something being shoved over into a new position, or of something having shifted, sometimes to release a blockage and sometimes to cause one.

Contributors

Alana Bartol comes from a long line of water witches. In her art practice, she explores concepts of visibility, transformation, and survival through our relationships with the non-human world and each other. Through performance, research-based, and community embedded practices, her site-responsive works propose dreaming, walking, and divination as ways of understanding across place, species and bodies. Her participatory works invite others to engage in acts of trust, inquiry, care, and improvisation, while making visible unseen forces that shape our world. Her work has been presented and screened nationally and internationally at various galleries including Plug In Institute of Contemporary Art (Winnipeg), ARC Gallery (Chicago), Karsh-Masson Gallery (Ottawa), Simultan Festival (Romania), Museo de la Ciudad (Mexico), Access Gallery (Vancouver), InterAccess (Toronto), and Groupe Intervention Vidéo (Montréal), among others. Recent residencies include The Banff Centre, Neighbourhood Time Exchange, and the Santa Fe Art Institute. She currently lives in Calgary and teaches at Alberta College of Art+Design. http://alanabartol.com/

IAIN BAXTER& is a Canadian conceptual artist, a teacher, and a thinker of art. He is one of the founders of Canadian conceptual art in the 1960s through the establishment of the conceptual art project and enterprise *N.E. Thing Co. Ltd.* His work has been featured in numerous national and international exhibitions and is included in various public collections in Canada and the USA, including the National Gallery of Canada, the Art Gallery of Ontario, Vancouver Art Gallery, and the Museum of Modern Art in New York. Most recently, his work was featured at the Sorbonne in Paris (2017) in an exhibition titled "The Power of &," and in a 50-year retrospective, "IAIN BAXTER&: 1958-2011," at the Museum of Contemporary Art Chicago and the Art Gallery of Ontario (Toronto) in 2012. His achievements have been recognized by many awards and prizes, including the Order of Canada (2002), the Order of Ontario (2004), the Governor General's Award in Visual and Media Arts (2004), the Molson Prize in the Arts (2005), and the Order of British Columbia (2007). He legally changed his name from "Baxter" in 2005—adding the "&" to signify, as Canadian curator David Moos puts it, a "non-authorial take on art production ... an unending collaboration with the viewer and the means to question the artist's role."

Warren Cariou was born into a family of mixed Métis and European ancestry in northern Saskatchewan, Canada. He has published works of fiction and memoir as well as critical writing about Indigenous storytelling, literature, and environmental philosophy. He has also created several photography and video projects about Indigenous communities in western Canada's tar sands region, and he has written numerous articles, stories, and poems about Indigeneity and petroleum. In his role as Director of the Centre for Creative Writing and Oral Culture at the University of Manitoba, he works with some of Canada's most celebrated Indigenous storytellers to promote cultural sovereignty through the performance of traditional narratives. He holds a Canada Research Chair at the

University of Manitoba and teaches in the Department of English, Film and Theatre. http://www.warrencariou.com/

Louise Chance-Baxter& is a visual artist and longtime collaborator with IAIN BAXTER&. Her paintings, while based on actual landscape and still life settings, are filtered through memory and, in translation, retain a freshness of visual that is often absent from treatments of similar subject matter. The artist's background in Trinidad-Tobago imbues the work with a tropical sensibility and her canvases exude warmth and humor, drawing the viewer into a world of the painter's imagination. Louise Chance-Baxter& has exhibited in Montreal, Toronto, Calgary, Windsor, Vancouver Island, Paris, Nice, Geneva, Xian and Shanghai and is in numerous private collections.

Raised on a farm in Wisconsin, Beth Franks had an idyllic childhood, a disadvantage for an aspiring writer. She survived the trauma of junior high by immersing herself in pulp fiction and nineteenth century novels. After graduating from UC Berkeley, she moved to London, England, where she and her husband lived for six years. Franks worked with disabled pupils, eventually teaching disability studies at Hobart and William Smith Colleges in New York. Currently she lives in California. Her poetry and fiction have been published in an eclectic range of journals from *Zyzzyva* to *Women's World.*

George Gessert is an artist and writer. He has exhibited widely in the United States, Europe, Canada, and Australia. His writings have appeared in many publications and have been translated into nine languages. *Green Light: Toward an Art of Evolution* was published by MIT Press in 2010. He was awarded a Pushcart Prize, and one of his essays was chosen by David Foster Wallace for inclusion in *Best American Essays*. Currently Gessert is working on *Growth*, a book of short stories that alternate between science fiction and the experience of having cancer. Three stories from *Growth* appear in *Naturally Postnatural.*

Ted Hiebert is a Canadian visual artist and theorist and Associate Professor of Interdisciplinary Art at the University of Washington Bothell. He is the author of *In Praise of Nonsense: Aesthetics, Uncertainty and Postmodern Identity* (McGill-Queens University Press, 2012), *A formalized forum for informal inquiry* (Noxious Sector Press, 2015), and (with David Cecchetto, Marc Couroux and Eldritch Priest) *Ludic Dreaming: How to Listen Away from Contemporary Technoculture* (Bloomsbury, 2017). http://www.tedhiebert.net

Christian Kuras lives in suburban Manchester, England. His work has been shown and published across Canada, the United States, and Europe.

Marta de Menezes is a Portuguese artist and a PhD candidate at the University of Leiden. De Menezes's practice explores the intersection between Art and Biology, working in research laboratories to demonstrate how new biological technologies can be used as an artistic medium. In 1999 de Menezes created her first biological artwork (*Nature?*) by modifying the wing patterns of live butterflies. Since then, she has used diverse biological techniques including: functional MRI of the brain to create portraits where the mind can be visualised (*Functional Portraits,* 2002); fluorescent DNA probes to create micro-sculptures in human cell nuclei (*nucleArt,* 2002); sculptures made of proteins (*Proteic Portrait,* 2002-2007), DNA (*Innercloud,* 2003; *The Family,* 2004) or incorporating live neurons (*Tree of Knowledge,* 2005) or bacteria (*Decon,* 2007). Her work has been presented internationally in exhibitions, articles and lectures. She is currently the artistic director of Ectopia, an experimental art laboratory in Lisbon, and Director of Cultivamos Cultura in the South of Portugal. http://martademenezes.com/

Natasha Myers is an Associate Professor in the Department of Anthropology at York University, the convenor of the Politics of Evidence Working Group, director of the Plant Studies Collaboratory, co-organizer of Toronto's Technoscience Salon, and co-founder of the Write2Know Project. Her book *Rendering Life Molecular: Models, Modelers, and Excitable Matter* (Duke University Press, 2015) is an ethnography of an interdisciplinary group of scientists who make living substance come to matter at the molecular scale. She has been working at the cusp of art, science, and anthropology for decades, first as a dancer/choreographer and plant scientist, and more recently as an anthropologist of art, science, and ecology. Her current projects span investigations of the arts and sciences of vegetal sensing and sentience, the politics and aesthetics of garden enclosures in a time of climate change, and most recently, she has launched a long-term ethnography experimenting with the arts of ecological attention in High Park's Oak Savannah. http://natashamyers.org

Kira O'Reilly is an artist currently based in Helsinki where she leads a pilot Masters programme in in Ecology and Contemporary Performance at the Theatre Academy, University of the Arts Helsinki. Her practice, both willfully interdisciplinary and entirely undisciplined, stems from a visual art background; it employs performance, drawing, makings, biotechnical practices, and writing with which to consider speculative reconfigurations around *The Body*. But she is no longer sure if she even does that anymore. She writes, teaches, mentors, and collaborates with humans of various kinds, technologies and non-humans of numerous divergences including mosses, spiders, the sun, pigs, cell cultures, horses, micro-organisms, bicycles, rivers, landscapes, tundras, rocks, trees, shoes, food, books, air, lichen, green glitter, moon, and ravens. A publication, *Kira O'Reilly: Untitled Bodies* edited by Harriet Curtis will be published during winter 2017 in the Intellect Live Series by Live Art Development Agency and Intellect Books. http://www.kiraoreilly.com/

Melentie Pandilovski is an art theorist and historian, curator, and art critic. He is Director of the Riddoch Art Gallery and Manager of Cultural Services with the City of Mount Gambier, Australia. He was previously Director of Video Pool Media Arts Centre (Winnipeg, Canada), the Visual Cultural Research Centre, Euro-Balkan Institute (Skopje, Macedonia), the Experimental Art Foundation (Adelaide, Australia), and the Soros Center for Contemporary Arts (Skopje, Macedonia). He has curated more than 150 exhibitions and organized numerous symposia, conferences, and workshops, in Europe, Australia, and Canada, including among others: "Age of Catastrophe" (2015),

"Toxicity" (2013/14), and "Marshall McLuhan & Vilém Flusser: Communication & Aesthetics Theories Revisited" (2012). His theoretical research examines the links between art, culture, technology, identity, and consciousness. He has published and edited a range of works on those topics, most recently, *The Rise of Bio-Society* (London: Palgrave Macmillan, 2018), "How biotechnology and society co-constitute each other," *Technoetic Arts* 10.1 (2012), and "On Modes of Consciousness(es) and Electronic Culture," *Glimpse—a Journal of Phenomenology and Media* v.2.

Paul Vanouse is a Professor of Art and Director of the Coalesce Center for Biological Art at the University at Buffalo. Interdisciplinarity and impassioned amateurism guide his art practice. His biological and interactive media projects have been exhibited in over 25 countries and widely across the US. His recent projects, "Latent Figure Protocol," "Ocular Revision," "Suspect Inversion Center," and "America Project" use molecular biology techniques to challenge "genome-hype" and to engage issues surrounding DNA fingerprinting, particularly the idea the most authoritative image of our time, the DNA fingerprint, is somehow natural. He has a BFA from the University at Buffalo and an MFA from Carnegie Mellon University. http://www.paulvanouse.com/

Amanda White is a PhD candidate in the Cultural Studies program at Queen's University and a visual artist based in Toronto. She has exhibited her artwork throughout Canada at galleries including the Harbourfront Centre, The Banff Centre for the Arts and Creativity, Plug In Institute of Contemporary Art, the Ontario Science Centre and Modern

Fuel Gallery, as well as independently producing many public interventions and engagements. Recent publications include articles for *esse art + opinions* (2016), *Antennae: The Journal of Nature in Visual Culture* (2017), as well as chapters for edited collections including; *Perma/Culture: Imagining Alternatives in an Age of Crisis* (Routledge, 2017) and *Why Look at Plants?* (Brill, forthcoming). http://amandawhite.com/

Robert Zwijnenberg is Professor of Art and Science Interactions at Leiden University, Faculty of Humanities. Trained in civil engineering and philosophy, he received a PhD in philosophy from the University of Amsterdam. His research and teaching focus on the role of contemporary art in academic and public debates on the ethical, political, aesthetic, judicial, and societal implications of the life sciences, with a focus on human enhancement technologies. In his research projects and academic courses he collaborates with bioartists who go into the lab to create art projects with the tools and materials of biotechnology, to explore biotechnology from an artistic and humanities perspective. Zwijnenberg is director of The Arts and Genomics Centre, a platform for stimulating, initiating, and supervising collaboration and exchange among international artists, genomics researchers, and life sciences professionals. http://www.artsgenomics.org/

Made in the USA
Las Vegas, NV
10 November 2026

57858720R00152